Ecoregion-Based Design for Sustainability

Springer
New York
Berlin
Heidelberg
Barcelona
Hong Kong
London
Milan
Paris
Singapore
Tokyo

Frontispiece. Summer rainstorm over the Great Plains in North Dakota. This is a common storm pattern in this ecoregion. Moist air masses from the Gulf of Mexico circulate upslope, bringing precipitation to the Plains during spring, summer, and fall. These storms are frequently accompanied by severe hailstorms, lightning, and occasional tornadoes. Photograph by John S. Shelton, reproduced with permission.

Robert G. Bailey

Ecoregion-Based Design for Sustainability

Illustrations by Lev Ropes

With 100 Illustrations, 93 in Full Color

Springer

Robert G. Bailey
Inventory and Monitoring Institute
USDA Forest Service
2150 Centre Avenue, Bldg. A
Fort Collins, CO 80526
USA
rgbailey@fs.fed.us

Cover Illustration: This house near Taos, New Mexico, is set in the matrix of the native
landscape, ensuring continuity and creating a definite sense of place. *House at Arroyo
Hondo*, watercolor, by Pam Furumo © 1987, reproduced with permission.

Library of Congress Cataloging-in-Publication Data
Bailey, Robert G., 1939–
    Ecoregion-based design for sustainability / Robert G. Bailey.
       p. cm.
    Includes bibliographical references (p.   ).
    ISBN 0-387-95429-5 (hardcover : alk. paper); ISBN 0-387-95430-9 (softcover : alk. paper)
    1. Landscape ecology.   2. Ecological landscape design.   3. Regional
planning—Environmental aspects.   I. Title.
QH541.15.L35 B35 2002
577—dc21   2001059796

ISBN 0-387-95429-5   (hardcover)   Printed on acid-free paper.
ISBN 0-387-95430-9   (softcover)   Printed on acid-free paper.

Printed in Spain.

9 8 7 6 5 4 3 2 1   SPIN 10864406   (hardcover)   SPIN 10864414   (softcover)

www.springer-ny.com

Springer-Verlag   New York Berlin Heidelberg
A member of BertelsmannSpringer Science+Business Media GmbH

For Susan Strawn

Preface

We are increasingly concerned, even alarmed, at the impact humans have on the Earth. With the geometric increase in population, concerns of sustainability are paramount. Yet many designers, land-use planners, hydrologists, developers, and others involved in this human impact continue to look at the environment piecemeal by focusing on the impact on a small area.

Design for sustainability requires us to look at the larger environment. Although we cannot correct many past mistakes, we can modify some and strive to design better.

My previous two books, *Ecosystem Geography* (Springer-Verlag, 1996) and *Ecoregions: The Ecosystem Geography of the Oceans and Continents* (Springer-Verlag, 1998), describe the vital importance of understanding regional ecology and regional-scale ecosystem units, or simply ecoregions. Many planners, heads of federal agencies, ecologists, and other professionals realize that we must consider the larger geographic scale of human impact in our land-management and conservation programs.

In this third book, I synthesize and illustrate the key principles of design and planning that relate to ecoregions and expand ecoregion concepts to include the human factors. This book explains the value of taking an ecoregional approach to sustainability. It completes my ecoregion triology and provides examples of the application of ecoregion concepts to professionals who design and manage our planet's beleaguered land and water resources. This book moves beyond defining ecoregions, to showing how an awareness of that definition can play a significant role in the search for sustainability.

In the course of my 35-year career as a Forest Service geographer, I have made research expeditions to many of the ecoregions of North America, Europe, Asia, and Africa. However, most of the examples I have included in this book are drawn from the regions I know best, namely those where I have lived and worked the longest: San Dimas Experimental Forest near Los Angeles (California Coast Range); Ogden, Utah (Intermountain Semi-Desert and Desert); Teton National Forest (Middle Rocky Mountains); Lake Tahoe Basin (Sierran Mountains); Fort Collins, Colorado (Great Plains Steppe).

This book is not a design handbook or technical reference filled with detailed case studies. Instead, it is designed to broaden the perspective of designers and planners so that they can address problems on a much larger spatial scale. As such, it is more philosophical than a how-to guide. Because the subject matter is inherently transdisciplinary, terms in bold are defined in the Glossary. Also included are a Resource Guide and a Selected Bibliography to assist those who take a deeper interest in the subject.

Fort Collins, Colorado   Robert G. Bailey
November 2001

Acknowledgments

A book is not the sole effort of the author, but is built from the work of others. I am indebted to other authors for their wonderful and insightful books on ecological design. First, I wish to acknowledge Sim Van der Ryn and his colleague Stuart Cowan at the University of California, Berkeley. Their remarkable book, *Ecological Design* (1996), provided inspiration to write my own book. Next, I want to mention the work of Wenche Dramstad and her colleagues at Harvard (*Landscape Ecology Principles in Landscape Architecture and Land-Use Planning*, 1996), as well as that of Dianna Lopez Barnett (*A Primer on Sustainable Building*, 1995). Finally, my admiration for Joan Woodward's *Waterstained Landscapes: Seeing and Shaping Regionally Distinctive Places* (2000) led me to see the connection between regions and design. I feel I owe much to that book, which I wish was available years ago. And, of course, my book builds upon the pioneering groundwork laid in Ian McHarg's *Design with Nature* (1969).

I would like to acknowledge Bob Alexander of the U.S. Geological Survey (now retired) for starting my thinking about sustainability at the ecoregional scale. I also wish to acknowledge two of my Forest Service colleagues, Eric Smith and Gordon Warrington (retired), who suggested ideas for inclusion in the book. Recognition for support should also go to Tom Hoekstra, Director of the Inventory and Monitoring Institute of the U.S. Forest Service. As always, it has been a pleasure to work with Lev Ropes of Guru Graphics and Linda Ropes of Creative Ink. The framework for this book is based on artwork that Lev prepared for lectures I have given on this subject. The text for the book is an extended caption for the artwork.

I wish finally to make mention of Susan Strawn. It is with deepest gratitude for her love and support, and all the time, talent, and ideas offered during the past 10 years, that I dedicate this book to her.

Fort Collins, Colorado   Robert G. Bailey
November 2001

Contents

leaving connections and corridors.   Honoring wide-scale ecological processes.

Local systems within the context of larger systems.   Spatial transferability of models.   Links between terrestrial and aquatic systems.   Design of sampling networks.   Landscaping and restoration.   Regional environmental problems.   Transfer knowledge.   Enhancing vegetation maps.

Current applications.   Future possible applications.

Spatial pattern matters.   Context is usually more important than content.   Matching development to the limits of the regions where we live.   What is the next step?

Appendix B   Climate Diagrams   161

Appendix C   Resource Guide   167

Appendix D   Common and Scientific Names   175

Appendix E   Conversion Factors 179

               Glossary of Terms as Used in This Book   181

               Notes   193

CHAPTER 1

Introduction

M any of today's buildings, construction practices, and land-use patterns are not sustainable. They do not work very well now, and they will not work at all in the long run. In *Ecological Design,* Van der Ryn and Cowan (1996, p. ix) state the following:

> If we are to create a sustainable world—one in which we are accountable to the needs of all future generations and all living creatures—we must recognize that our present forms of agriculture, architecture, engineering, and technology are deeply flawed. To create a sustainable world, we must transform these practices. We must infuse the design of products, buildings, and landscapes with a rich and detailed understanding of ecology.

This statement parallels Wann's observation in his book *Biologic* (1994, pp. xvi, 3) "that environmental deterioration is a lack of relevant information . . . [and that] poor design is responsible for many, if not most, of our environmental problems." Much of this situation has come about at a time when complacency and consumption have overshadowed human connections to natural systems.

Sustainable design in human developments has come to the forefront in the last 20 years. It is a concept that recognizes that human civilization is an integral part of the natural world and that nature must be preserved and perpetuated if the human community is to sustain itself indefinitely. Sustainable design is the philosophy that human development should exemplify the principles of conservation and encourage the application of those principles in our daily lives.

In order to integrate ecology and design, we must mirror nature's deep interconnections with our own way of thinking about design. The concept of sustainable design holds that future technologies must function with the way nature works. Unfortunately, many planners and managers continue to look at the environment piecemeal by focusing only on local impact of development. We must consider human impact on large geographic scales in land management and conservation programs.

Understanding Whole Systems

The good news is that, in the United States, the Bureau of Land Management, Fish and Wildlife Service, Forest Service, and National Park Service are phasing into a radically new approach to managing the public lands. They are shifting from their focus on individual resources, such as timber, to a more holistic approach of managing *whole* **ecosystems**.[1] In simple terms, the ecosystem concept states that the earth operates as a series of interrelated systems within which all components are linked, so that a change in any one component may bring about some corresponding change in the other components, and in the operation of the whole system. This new ecosystem approach to land management stresses the interrelationships among components and how they are combined (or integrated) rather than treating each one as a separate characteristic of the **landscape** (Fig. 1.1). This approach provides a basis for making predictions about resource interaction, such as the effects of timber harvesting on water quality.

One method of capturing this integration is the **ecological land classification** technique (Rowe and Sheard 1981). This technique includes the delineation of units of land that display similarity among a number of components, particularly in ways that may affect their response to management and resource production capability. How these components are integrated can be shown at two general levels (at different scales). One level shows the integration within the local area, and another shows how the local area is integrated and linked with other areas across the landscape to form larger ecosystems.

There are several reasons for recognizing ecosystems at various scales. Because of the linkage between ecosystems, a modification of one ecosystem may affect the function of surrounding ecosystems. Furthermore, how an ecosystem will respond to management is partially determined by relationships with surrounding systems linked in terms of runoff, groundwater movement, microclimate influences, and sediment transport.

Multiscale analysis of ecosystems pertains to all kinds of land. Many planning issues transcend ownership and administrative boundaries and are multiagency, multistate, and international. These issues include air pollution, **anadromous fisheries**, forest insects and disease, and biodiversity. To address these issues, the planner must consider how geographically related ecosystems are linked to form larger ecosys-

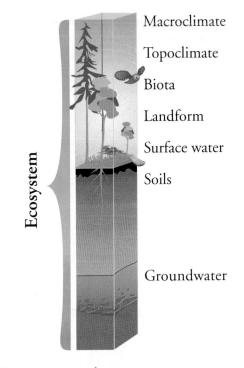

Macroclimate

Topoclimate

Biota

Landform

Surface water

Soils

Groundwater

Ecosystem

Figure 1.1. The components of an ecosystem.

tems, regardless of ownership. This will also require government scientists and researchers to integrate their efforts across agency lines.

For example, in the Forest Service's Natural Resource Agenda, where the focus is to sustain and restore the fabric of whole ecosystems, the ownership and political boundaries are logically giving way to boundaries set by ecological criteria. The Forest Service is beginning to look beyond national forest boundaries and expand its horizons to view the forests from a larger ecosystem-based perspective (Fig. 1.2).

The Forest Service and other organizations are beginning to look at whole **ecoregions**: large, regional-scale ecosystems, like the Laurentian Mixed Forest and the Sonoran Desert.[2] Managing and conserving relatively small parcels of land will not do the job alone. What sustains ecosystems is large functional landscapes, where wildlife can migrate and respond to natural processes—like fire. We need a larger vision, and that is what ecoregions give us. We will continue to manage ecosystems at the local level, but we need to do so within the larger ecoregional context. We need to work with an array of partners and to provide greater involvement with local communities. This is necessary because ecoregions are not contained by political borders or administrative boundaries.

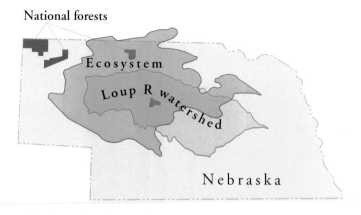

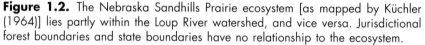

Figure 1.2. The Nebraska Sandhills Prairie ecosystem [as mapped by Küchler (1964)] lies partly within the Loup River watershed, and vice versa. Jurisdictional forest boundaries and state boundaries have no relationship to the ecosystem.

The need for a regionalism has been recognized for many years. In December 1935, the National Resources Committee's report, Regional Factors in National Planning and Development, recognized that some national problems—notably use of natural resources—would require approaches based on boundaries that extend beyond conventional political jurisdictions (Foster 1997). Others outside the government have played their part in influencing this regionalism, attachment to place, and environmentalism, such as artists and illustrators like William Henry Holmes, Thomas Hart Benton, Grant Wood, Maynard Dixon, and Georgia O'Keeffe, and writers like Wallace Stegner, John Brinckerhoff Jackson, and Ann Zwinger. Readers who seek an impression of the art movement known as **Regionalism**, in which Grant Wood (1892–1942) was a seminal figure, will find it in Wanda Corn's catalog of a traveling exhibition, *Grant Wood: The Regionalist Vision* (1983). In *Young Corn*, Wood captures wonderfully the Prairie Parkland ecoregion—characterized by intermingled prairie (replaced by corn fields), groves, and strips of deciduous trees—around Cedar Rapids, Iowa (Fig. 1.3).

Sustainable Design

The kinds of regional ecosystems vary vastly, including their ability to sustain impact. A trail in the tropical rainforest might disappear after a few years, but in the semiarid brushlands of Wyoming, it might take a century or more. The Mormon Trail, for example, is still visible 140 years after it was traveled (Fig. 1.4).

Figure 1.3. *Young Corn,* painted by Grant Wood in 1931. Oil on masonite panel, $23\frac{1}{2} \times 29\frac{7}{8}$ inches. Copyright Cedar Rapids, Iowa, Community School District. Memorial to Linnie Schloeman, Woodrow Wilson School.

Optimal management of land ensures that all land uses consistently sustain resource productivity *and* maintain ecosystem processes and function. This equals ecosystem capability, and capability provides the context for looking at land-management options. The expression for this relationship is

Sustainability = Resource productivity
                              + ecosystem maintenance = Capability

Sustainable design is the process of prescribing compatible land uses based on the limits of a place, locally as well as regionally. Sustainability, the capability of the natural systems to maintain themselves

Figure 1.4. Mormon Trail ruts cut into sagebrush steppe near Muddy Gap, Wyoming. Photograph by Lev Ropes.

while still being used, is the key, whether it is called designing for sustainability, sustainable development, design with nature, environmentally sensitive design, or holistic resource management.

The Need for Regional Ecology

The need for a **regional ecology** approach is clear. When human intervention into ecological processes was small, nature often compensated. Furthermore, many impacts were localized, such as small-scale grazing in a river **flood plain** (Fig. 1.5). This is not the case anymore, as evidenced by human disturbance of wild vegetation of whole continents (Fig. 1.6).

During the nineteenth century, settlers tamed the eastern half of the continent, cutting forests, plowing prairies, and draining swamps. The United States government exiled native peoples such as the Cherokees and Creeks west of the Mississippi. Wildlife declined everywhere, and several species became extinct, notably the passenger pigeon. In the

Figure 1.5. Cattle roundup in the American West, circa 1900. Vintage postcard by Detroit Publishing Co.; author's collection.

Far West, the influx of Europeans had a devastating effect on the indigenous people and the environment. The California gold rush in the mid-1800s resulted in the arrival of miners in the thousands, who laid hills bare, and turned streams into muddy trenches. Many disappointed miners became farmers, turning the vast seas of grass into agricultural land and felling huge swaths of timber for housing and railroads (Fig. 1.7). Herds of elk, pronghorns, and big horn sheep that had once been

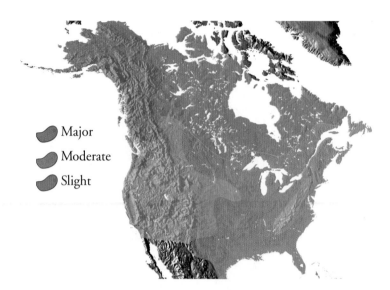

Figure 1.6. Human disturbance of wild vegetation of North America. Redrawn from *Man and Earth's Ecosystems*, by Charles F. Bennett (1975).

Figure 1.7. Logging redwoods in a California forest in the 1890s. Vintage postcard by Pacific Novelty Co., San Francisco; author's collection.

abundant everywhere disappeared within decades, ranchers replacing them with sheep and cattle. The ranchers saw wildlife as competition to be eliminated. Predators such as wolves, mountain lions, and coyotes were prime targets.

In the Great Plains, large areas of natural grassland have disappeared, replaced by farmland. This is the result of the 1862 Homestead Act that gave pioneer families 160 acres (65 hectares) of free federal land if they could farm it for five years. These are America's steppes— windswept, nearly treeless, and largely semiarid—that are so dry and desolate that pioneers called it the Great American Desert. This process of transformation was aided by the large number of small-town blacksmith–inventors in the Midwest. Most noteworthy is John Deere, who invented the steel plow. With this new technology, farmers could break the thick sod. The invention of barbed wire in 1873 made possible the inexpensive fencing of farms, which kept cattle drives and wildlife out of cultivated areas (and also changed the livestock industry).[3] Windmills made it possible for them to draw water to the surface for their livestock and farm animals. Thus, because water was available for domestic use as well as animals, human settlement in isolated areas became possible. The expansion of American civilization across the Plains also was shaped by another technological advance—the railroads. The railroads dictated where the rail went and, in the process, the locations of the farming communities. They were the mechanism that brought settlers and goods into the Plains and exported crops out. Without a railroad town, there could be little commerce to serve the growing population. In many cases, the rails were laid first and then the towns were situated at intervals along the rail corridor. The railroad companies offered incentives and made exaggerated claims about the advantages of a particular location in order to entice homesteaders. This same kind of boosterism was later used by the railroads and land developers to get people to move to southern California at the beginning of the twentieth century. The Southern Pacific Railroad Company started *Sunset* magazine in 1898, which continues publication to this day, as a sophisticated marketing tool to advertise the weather, scenery, and lifestyle of the region.

The present large-scale effects of human impact need to be evaluated on a macroscale. This is not only due to their large spatial nature [e.g., conversion of natural areas to crop monocultures such as farmland near Fort Collins, Colorado (Fig. 1.8)], but also because of the **cumulative effects** from aggregated local activities such as urban sprawl (Fig. 1.9). Furthermore, leapfrogging, sprawling development erodes existing communities, converts prime farmland to housing, requires expensive new highways for commuters, reduces biodiversity, and harms the environment.

Furthermore, we must discard the artificial approach of outlining re-

Figure 1.8. Short-grass steppe converted to farmland near Fort Collins, Colorado. Author's photograph.

gional ecology problems that are strictly terrestrial or aquatic. Ecosystems related by geography are not necessarily related by common properties. An area of spruce forests and glacially scoured lakes, for example, constitutes a single ecosystem that is linked by downhill flows of water and nutrients, through coarse Spodosol soils,[4] toward clear oligotropic ("few foods") lakes (Fig. 1.10). Geographical related systems such as this, unified by an exchange of energy and material, may be combined into larger geographic units referred to as a "**landscape ecosystem**." Aside from interconnections between these ecosystems, the role of humans at regional scales begs against such simplistic considerations.

We simply cannot regard terrestrial and aquatic components of landscape ecosystems as independent ecosystems, because they cannot exist apart from one another. Just as the lower part of a slope exists only in association with the upper, gullies could not form if no **watershed** existed. The units of a landscape always comprise connected or associated ecosystems. As stated earlier, within such an ecosystem the diverse component ecosystems are mutually associated into a whole by

Figure 1.9. Suburban development ignores the natural environment as it sprawls across Jefferson County, Colorado, part of the southwest Denver metropolitan area. Foreground reveals the subdivision's semiarid context. Photograph by Lev Ropes.

the process of runoff and the migration of chemical elements. Their common history of development also unites them. Streams are dependent on the terrestrial systems in which they are embedded. Therefore, they have many characteristics in common within a given terrestrial system, including biota and hydrology. What we do to the terrestrial directly affects the aquatic. The most obvious example is sediment and woody debris deposited in streams and lakes as a result of increased erosion of the slopes surrounding the water body (Fig. 1.11). Sediment tends to have a negative effect, whereas the woody material is remarkable, as it helps host and nurture life for multiple communities (Maser and Sedell 1994).

The reverse situation is also true, as impoundments lead to channel aggradation upstream and channel deepening downstream, among other problems (Fig. 1.12). An impoundment such as a dam creates an artificial **base level** for the stream or river upstream from the dam. As a result, its erosive power is decreased and its ability to carry sediment is reduced, so sediment is deposited at the upstream end of the reservoir. Because of this, the water released from the reservoir, with little

Figure 1.10. Spruce forests and glacially scoured lakes in Voyageurs National Park, Minnesota. Photograph by Jack Boucher, National Park Service.

or no sediment, has an increased ability to cut the channel deeper below the dam. Aggradation of the main channel above the dam and degradation below the dam affects the channels of tributaries flowing into the main channel because they are flowing to a new base level. The dam, therefore, has a regional effect far beyond the immediate vicinity of the structure and will differ considerably depending on the regional ecosystem in which it lies.

Such changes are not restricted to channel characteristics alone. Pringle (1997) addresses the biological changes that can occur in headwater systems as a result of downstream habitat deterioration and hydrologic modifications. Such changes include reduced genetic flow and variation in isolated populations because species no longer migrate upstream. This changes the regional patterns of biodiversity. The current

Figure 1.11. Sediment from Incline Creek, Nevada flowing into Lake Tahoe. Unknown photographer.

controversies over salmon spawning in the Columbia River basin are a prime example.

In the construction of dams in the western United States, almost no attention was paid to the ecological aspects of such hydrological modification, and even today these aspects have not been fully researched. It was discovered rather soon that the water temperature downstream from a dam is often greatly altered—usually lowered—and that the ecology of the water trapped above a dam is markedly different from before and supports different biota than had previously existed.

However, the dams were not isolated constructions. They were tied to areas often far away from their sites. For example, water is conveyed by tunnel and aqueduct to Los Angeles and San Francisco from dammed sources far away. Water from the Colorado River irrigates citrus groves hundreds of kilometers away (Fig. 1.13). The western United

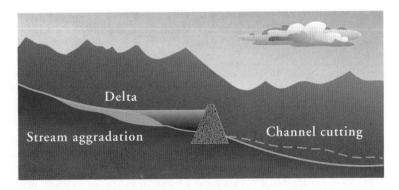

Figure 1.12. A dam affects the river channels upstream and downstream for considerable distances.

Figure 1.13. Boulder Dam (renamed Hoover Dam in 1947) rises 70 stories from the bed of the Colorado River. Although the dam appears minuscule compared to Lake Mead, whose length is greater than 160 km, the dam may outlast the reservoir due to siltation. Vintage travel brochure by Union Pacific Railroad, 1941; author's collection.

States is now largely a **hydraulic culture** dependent on complex modifications of the surface and subsurface waters. As noted by Bennett (1975), these hydrological modifications of the West are the most striking aspect of human modification of western ecosystems, except for the urbanized areas in Southern California and those adjacent to the San Francisco Bay. The economics and politics behind these modifications are provided by Reisner in his book *Cadillac Desert* (1986).

Dams are just one of many ways in which management decisions may reach far beyond the initial goals. The following chapters give further examples.

Rationale for Land Management in an Ecoregional Context

Historically, the ecosystem has been defined as a small homogeneous area, or **site**, such as a stand of trees or a meadow. However, as stated earlier, ecosystems occur on many scales, from small sites to ecosystems of regional scale. The regional approach is much more useful for planning and management than our traditional scattered, small-scale analysis because all ecosystems operate within a context of larger ecosystems (Fig. 1.14).

An illustration of this idea that I especially like comes from the nineteenth-century Japanese painter and print designer, Katsushika Hokusai (1760–1849), which portrays a great ocean wave. *The Great Wave off Kanagawa* (circa 1830) shows a large wind-blown wave with Mt. Fuji in the background. Contained within the scene are similar waves at many levels of scale or size. It may not have been the artist's intention, but his world-famous rendering is a beautiful illustration of small systems nested within a larger system.

The waves of the oceans are part of a larger ocean ecosystem that, in turn, controls other systems. For example—as fog along the California shows (Fig. 1.15)—the large ocean systems control the ecosystem patterns of the continents that are embedded in the oceans through their influence on climatic patterns. From our knowledge of the larger ecosystems, we can much better understand the smaller subsystems

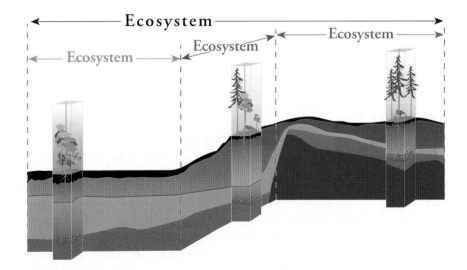

Figure 1.14. Ecosystems are nested or reside within each other.

Figure 1.15. Fog uplift from summit of Mt. Tamalpais, California. Vintage post-card by Pacific Novelty Co., San Francisco; author's collection.

and predict the outcome of land use and/or natural resource development. Thus, the context that the larger system provides is very important. For example, a **meadow** embedded in a forest operates much differently than a large expanse of grassland because of interaction with the surrounding forest.

The Gap

Saying that we need to manage ecosystems is one thing; actually doing it is another. There is a gap between what we as ecosystem analysts and managers say and what we do. Part of the difficulty in implementing this approach is that disciplinary specialization is inherently incompatible with ecosystem management. The various disciplines still cling to studying ecosystems in terms of components: soil, vegetation, water, wildlife, and so on. This mindset sees the environment as composed of pieces, and the result is management from individual resources such as timber and wildlife rather than managing ecosystems as wholes. If we study the ecosystem only as broken into components, we cannot assemble the whole with any success. The Humpty Dumpty nursery rhyme is a metaphor for this traditional mindset (Fig. 1.16). The same applies to management of landscapes and regions composed of mosaics of small, local ecosystems. If we try to manage them individually without considering how they interact and influence one another, we will miss the wider ecological consequences of our designs.

Figure 1.16. "Humpty Dumpty sat on a wall. Humpty Dumpty had a great fall. All the king's horses and all the king's men couldn't put Humpty Dumpty together again." Mother Goose nursery rhyme; author unknown.

There is a unity and orderliness in nature. It is discernable and amenable to analysis. For example, all steep slopes have shallow soils. Recognition of this orderliness and unity is the basic tenet of ecosystem analysis.

Given this holistic view of nature, I see a need for evolutionary change in educational institutions to reduce compartmentalization of academic disciplines. Reductionism still holds sway in science. Traditionally trained scientists are taught to turn up the power on their microscopes to get a better look. What they really need to do is get a "macroscope" to get an integrated view. As noted by Ian McHarg (1997), integration requires bridging between separate sciences, an attitude resisted by universities and government institutions. Meteorologists study weather, geologists study rocks, hydrologists address water and watersheds, and pedologists focus on soils. An associated institution exists for almost every discipline. The U.S. Geological Survey controls rocks and water, soils repose in the USDA Natural Resources Conservation Service, and fish and wildlife are relegated to the U.S. Fish and Wildlife Service. The U.S. Forest Service is charged with managing our national forests and the U.S. Bureau of Land Management fills a compatible role for rangelands. In many of these agencies, problems of integration are compounded by the separation of functions. Commonly they have a separate staff set up for wildlife and fisheries, a staff for timber, and a staff for recreation, each with its own budget. No common unifying framework has existed to bring the functions together.

This can happen if we lose sight of the complete, comprehensive ecosystem. A photograph printed in a newspaper helps illustrate this point. When the photograph is significantly enlarged, it appears as groups of dots. The same thing happens if a painting by the French impressionist who invented pointillism, Georges Seurat, is viewed too closely. Only when the image is viewed from a distance does the nature of the image become comprehensible. This is a good metaphor for the type of complications that can result when landscapes and ecosystems are examined too closely. In his master's thesis, Isaac Brewer (1999) observes that music can further aid in the explanation of this point. Mozart created compositions comprising a myriad of individual notes. The true nature of the musical score cannot be understood by knowing the pitch and duration of each individual note. The individual notes do not impart the listener with the complete package of rhythm, melody, harmony, and tone inherent in musical compositions. Characters that are used in writing the Chinese language further illustrate the need for understanding context. There are over 8 thousand of these characters; their meaning depends on the character(s) with which they are associated. For example, when the character for flower (花; phonetic spelling, *hua*) is combined with the character for tea (茶; phonetic spelling, *cha*), we have *jasmine tea* (花茶); when the same character for flower is combined with the character for bed (坛; phonetic spelling, *tan*), we get *flowerbed* (花坛). Landscapes, also, cannot be understood by examining the individual parts; the entire region must also be investigated.

Holistic studies require a scientist to examine systems as *more than* the sum of its parts, as opposed to the narrow focus of one piece of the system. As we will see later, a system has properties that cannot be observed from simply looking at the pieces. Furthermore, a central tenet of holistic landscape ecosystem studies must always be the way the smaller systems fit together to form larger systems, or regions. Therefore, each of the areas surrounding a study area must also be taken into consideration.

It is simply not enough for each discipline to conduct inventories of the separate ecosystem pieces, write up what they found, and then have someone staple the pieces together and think that we have captured the integrated nature of the ecosystem. Stapling pieces together into a report—or overlaying maps with a geographic information system (GIS)—can not serve that function. We, as analysts, are the only ones capable of that job. An inventory of the pieces or components of an ecosystem provides an inventory of its anatomy; it does not necessarily provide an understanding of how the parts fit together (its structure or geometry) and function. As Rowe (1980, p. 44) points out, "The key to capturing the integrated nature of ecosystems is not to be found simply in the vegetation, in the soil profile, in the topography and geology, in the rainfall and temperature regimes, in the water bodies, but rather in the spatial coincidences, patterning and relationships of these functional components." In the next chapter, we consider the relationships of these components as the basis for an integrated approach to classifying land as ecosystems.

CHAPTER 2

Nature's Geometry

As we said earlier, ecosystems exist in different scales nested within each other (Fig. 2.1). The boundaries are open and permeable, leading to the transfer of energy and materials to and from other ecosystems. The open nature of ecosystem boundaries is important, for even though we may be dealing with a particular ecosystem as a land unit, we must keep in mind that the exchange of material with its surroundings is an important aspect of the system's function.

Because of these linkages, the modification of one ecosystem affects surrounding ecosystems, sometimes adversely. Furthermore, how a system will respond to management is partially determined by relationships with surrounding systems linked in terms of runoff, groundwater movement, microclimatic influences, and sediment transport (Chapter 1). Ecosystems do not exist in isolation. As discussed earlier, for example, the climate in a meadow is altered by the surrounding forest. We need to work on understanding these linkages so we can better predict the impacts of our activities.

We must examine the relationship between ecosystems of different scales in order to analyze the effects of management. A disturbance to a larger ecosystem will affect smaller-component ecosystems within. For example, logging on the upper slopes of an ecological unit will affect smaller systems downstream, such as stream and **riparian** habitats (Fig. 2.2). Other forms of vegetation manipulation may have similar effects. For example, **chaparral** species have deep root systems and therefore use more water than shallower-rooted grass species. Researchers found that converting from chaparral to grass to increase the water yield of steep experimental watersheds in southern California affected stream systems through increased debris production and discharge rates (Orme and Bailey 1971). As the roots of the deep-rooted chaparral species decay, they can no longer anchor the soil on steep

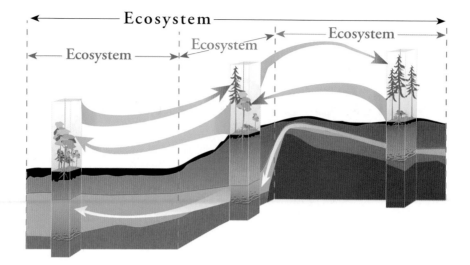

Figure 2.1. Ecosystems are nested with permeable boundaries.

slopes. The change decreased the stability of the slopes during storms and increased the amount of material washing downslope. Increased erosion is followed by severe gullying, which, in turn, is accompanied by aggradation of the main valley. We can extend this concept of scale-linking all the way from the small watershed to the whole planet. Therefore, a multiscale approach is much more useful than our traditional scattered, small-scale analysis because all systems function within a context of larger systems.

Because ecosystems are spatial systems and are consistently inserted, or nested, into each other, each level subsumes the environment of the system at the level below it. Therefore, it conditions or controls the behavior of the system at the level below it. For example, climate controls runoff in a watershed, which, in turn, interacts with hill slopes to produce stream channels. At each level, new processes emerge that were not present or evident at the next level. As Odum (1977) noted, research results at any level aid the study of the next higher level, but never completely explain the phenomena occurring at that level, which itself must be studied to complete the picture.

Some of the processes involved in a landscape composed of a mosaic of ecosystems may be in addition to those in its separate component ecosystem. They include those processes of interaction among the component ecosystems. For example, a snow–forest landscape includes dark pines that convert solar radiation into sensible heat that moves to the snow cover and melts it faster than would happen in either a

Figure 2.2. A meadow surrounded by forest in central Idaho. Photograph by U.S. Forest Service.

wholly snow-covered or wholly forested basin. The pines are the intermediaries that speed up the process and affect the timing of the water runoff. Watershed managers attempt to produce the same effect by strip-cutting extensive forests.

Linking Across Scales

Understanding links between ecosystems at various scales allows us to analyze **cumulative effects** of action at one scale and its effects at another. Scale-linking reminds us of the wider environmental consequences of our designs.

An example of a smaller ecosystem in a larger controlling system is the meadow embedded in a forest (Fig. 2.2). It will function differently than a large expanse of grassland (Fig. 2.3). The forest affects the microclimate and the plant cover of the meadow, sheltering the meadow from drying winds or from hail. Many bird species that nest in the forest feed in the meadow, and meadow rodents like to hibernate at the edge of the forest or in the interior.

At the zones of contact, or **ecotones**, between forest and meadow, the greatest concentration of animal life, mostly insects and birds, oc-

Figure 2.3. Short-grass steppe near Laramie, Wyoming. Author's photograph.

curs at the edge of the forest. This accounts for the higher density of animal populations in a forest–meadow landscape than in a forest land-scape or a grassland landscape (Odum 1971).

A single row of corn further helps to illustrate the effect of scale. This row, planted by itself, will not grow tall and may not produce ears of corn. On the other hand, the same corn plants growing in the center of a field of many plants of the same species will behave just the opposite. At the edge of the field, the heights of the rows become shorter and the plants less productive.

If we focus on a single scale, we miss the other scales, and hence miss opportunities to work across them in a unified way to address the problem. The migration of birds demonstrates how this works. The Swainson's hawk is a **neotropical migrant bird** that escapes winter by flying more than 22,500 km along the Pacific flyway from the dry steppes of California to similar steppes in Argentina. Researchers in California noted a decline in the population of hawks but were unable to explain it until they examined the problem at the global scale. In Argentina, they found the birds being killed in very large numbers by pesticides that were used by farmers of alfalfa and other grains to pro-tect their crops from pests. Scale-linking reminds us of the wider en-vironmental consequences of our designs.

Characteristically, conventional design tends to work at one scale at a time. Ecological design integrates design across multiple levels of scale, reflecting the influence of larger scales on smaller scales and smaller on larger. An overview of ecology from a scale-linking perspective is provided by Allen and Hoekstra (1992).

An Organizing Principle for Ecologically Compatible Design

In nature, geometry or structure (how the parts fit together) reflects underlying processes. Climatic and geologic processes operate to produce a variety of **landforms** (geology and topography). Vegetation responds to the landform features, with each plant community favoring a particular microclimate and set of soil conditions. Vegetation, in turn, is a major determinant of ecosystem structure and animal habitat. An example of this is the Canadian forest **formation**, schematically represented in Figure 2.4. This forest is dominated by the continental subarctic macroclimate with its long and severe winters and no warm season. If the landforms of this region were uniform, the vegetation of the whole climatic region would be comprised of a needleleaf evergreen forest in various stages of succession. However, the landforms are diverse and interspersed with swamps and lakes left by departing ice sheets. In some places, ice scoured the rock surfaces bare, entirely stripping off the overburden. Elsewhere, rock basins were formed and stream courses dammed, creating countless lakes and wetlands. Thus, the total vegetation cover is actually a mosaic of small units reflecting the inequalities in conditions of slope, drainage, and soil type. The nature of the landform, therefore, will allow a differentiation of many **habitats**, and the percentage of the region occupied by the ecosystem which characterizes the formation (needleleaf evergreen forest) will de-

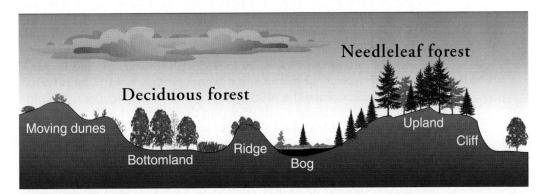

Figure 2.4. Geometry of the Canadian forest formation. Redrawn from Dansereau (1957).

pend on the amount of well-drained upland. Other ecosystems will occupy sites that are either dryer or wetter, forming a complex mosaic of ecosystems across the landscape.

The spatial context (i.e., nature's geometry) is an important organizing principle for **ecological design**. Van der Ryn and Cowan (1996, p. 18) defined ecological design as a "design that minimizes environmentally destructive impacts by integrating itself with living processes." It determines the context for design, both at the scale of small ecosystems and for whole regions. Over a century ago, Major John Wesley Powell (1834–1902), head of the U.S. Geological Survey, explicitly recognized this organizing principle in his suggestion to settle the arid West in a way that matched land allocations to the availability of water (Stegner 1992).

This idea of matching land use to the limits of the land had precedence with other cultures. As a result of fieldwork in South America, geographer Isaiah Bowman (1916) was stuck by the close correlation between topographic features and activities of man. In his study, he described Peru as having two broad regions: the Maritime Andes and the Eastern Andes. Within these regions, he described the dominant "topographic types" and their relationship to agriculture, settlement, pastoral life, and communications between groups. These topographic types represented such recurring features as high plateaus, basins, canyons, snow-covered mountains, and so forth. He used several techniques to illustrate these relationships, the most effective of which was the "regional diagram," as shown in Figure 2.5. Bowman's classifications were based on an important principle. A given composition of the biophysical elements of the earth is repeated essentially from place to place over a region. An important feature of Bowman's principle is its restriction to a given region, because elements that are related to each other in one region may be related to each other in a different way in another region. For example, ocean winds may produce rain and productive agriculture on the one side of a mountain range and leave a desert on the other.

Bowman's ideas were the basis for the design of the Michigan Economic Land Survey that was started in 1922. The integrated survey of soil, vegetation, farm use, and so forth that followed resulted in combinations of associated features that were call the "natural land type." During the years since the Michigan project, it has become the prototype for resource management planning throughout the United States, notable among which were the various Tennessee Valley Authority, Resettlement Administration, and Soil Conservation Service surveys. Today, these associated features are termed "ecosystems" and the regions where one finds the same associations is termed an "ecoregion."

A good example of spatial context is found in Joan Woodward's *Waterstained Landscapes* (2000), a book about regional patterns and

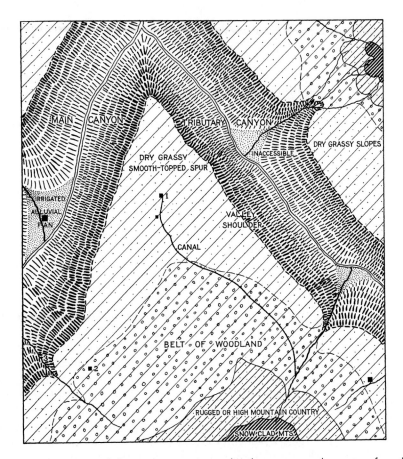

Figure 2.5. Regional diagram representing the deep-canyoned country of southern Peru. The dark hatchures represent the canyon type, where settlements are located on the only agricultural land, the open diagonal crosshatch represents the plateau type, used for pasture, and the close diagonal crosshatch represent the mountain type. From Bowman (1916), p. 58. Copyright © 1916 by The American Geographical Society; reprinted with permission.

process in the Great Plains steppe that lies between the Rocky Mountains and the 104th meridian, from the Canadian border through Oklahoma. The climate is semiarid with accompanying short grasses and shrubs covering the rolling to flat plains. Flowing out of the adjacent mountains and across the dry plains are widely spaced **exotic rivers** that derive their discharge from the mountains where water surplus exists. They are lined by **riparian** forests. The riparian forest hugs the river and is bordered by steppe grasslands beginning at the edge of the floodplain. The forests that line these rivers usually flood in spring when meltwater brings the river to crest. The floods are followed by summer drought, when evaporation tends to exceed precipitation, and the water level drops. Because of this annual cycle, western riparian forests tend to have broad, fertile floodplains, where sediment is deposited as waters recede. Underlying these floodplains are alluvial **aquifers** up to 6 m in depth.

Cottonwoods are one of the dominant trees of the riparian forest, because they find water and soil to support their needs. Their roots tap into the aquifer, protected from the evaporative forces of the semiarid climate. Planting these same trees on fine-grained plains soils, with only atmospheric precipitation to sustain them, would kill the tree. Feeding them water through irrigation would sustain them but would also change the region's water budget by distributing stored water in a highly evaporative condition. According to Woodward (2000, pp. 52–53), "Knowing the formative processes at work and adapting them would enable planting designers to create fitting designs requiring less maintenance and stirring up fewer problems down the road."

Knowing the **formative processes** also has a key role to play in assessing possible impacts on the environment that might be caused by new developments. Determining the likely effects of major projects on the natural environment through an environmental impact assessment has become a statutory requirement in many countries throughout the world. The probable effect of construction of dams and **tidal barrages**, for example, have been at the center of many large investigations. Some years ago, there was proposal to construct a tidal barrage (for electricity generation) in the estuary of the River Severn, between Wales and England. The distribution of **estuarine invertebrates** is determined largely by salinity levels. A tidal barrage would change the salinity distribution, which would affect the distribution of estuarine invertebrates. This, in turn, would affect the distribution of estuarine birds that feed on the invertebrates (Spellerberg and Sawyer 1999).

Out of Context

Current political and planning boundaries do not reflect the overlying structures and flow of ecosystems. The strip of cleared land along the border of Yellowstone National Park obviously does not delineate a natural ecological boundary (Fig. 2.6). In this case, this occurred because the land-management agencies have disparate missions and user groups. In Yellowstone National Park, timber harvesting is prohibited. And in the Targhee National Forest, large areas of trees were removed through **clear-cutting**.

Adjacent land practices and boundaries may block or disrupt vital ecological flows across large portions of a landscape. In their book

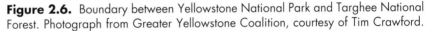

Figure 2.6. Boundary between Yellowstone National Park and Targhee National Forest. Photograph from Greater Yellowstone Coalition, courtesy of Tim Crawford.

Stewardship Across Boundaries Knight and Landres (1998) addressed the complex biological and sociological impacts of these boundaries. For example, irrigation diversion outside Everglades National Park altered hydrologic flow within the park, severely altering vegetation and wildlife communities. In contrast to disrupting flow, different management goals on one side of the border may promote or enhance the spread of certain ecological flows that are detrimental to the other side of the border. Fire management illustrates the problems of both disrupted and enhanced flows across borders. In the first case, fires that initiated in lower elevation areas and subsequently spread up into higher vegetation types are now largely suppressed, thereby also stopping the fires that would naturally have burned up into a higher elevation. In the second case, different management actions on one side of a border may result in fires spreading into areas that they typically would not reach.

Administrative boundaries affect wildlife populations in several ways, especially by disrupting dispersal, by fragmenting habitat into isolated areas, and by surrounding the isolated areas with conditions that are detrimental to survival. For example, for large herbivores such as elk that migrate seasonally, boundaries and adjacent land activities may prevent the seasonal migrations necessary for survival. Similarly, bison leaving Yellowstone National Park are killed because they may be vectors for brucellosis, a disease causing higher abortion rates in domestic livestock.

The boundary impacts may persist long after the boundary has served its intended purpose. The species composition on two sides of a tumbledown fence that once separated a pasture from a row-crop field is one example; an ancient Anasazi village boundary in the Colorado

Plateau is another. In the Great Plains, early settlers lined fields and planted trees around their farmhouses to block winds and reduce soil erosion. The Timber Culture Act of 1873 encouraged forestation by allowing patent to 160 acres (65 hectares) to any settler planting and maintaining 40 acres (17 hectares) of timber in treeless sections of the West. As a result, a number of tree claims were planted between 1873 and 1900. Many of these trees were imported from similar arid steppelike environments. The use and ecological significance of such contrasting patches may differ because of contrasting nutrient concentrations or vegetation composition and structure that have persisted into the present. In other cases, contrasts persist because of continuous heavy grazing: Soil texture and permeability changes occur as a result of repeated trampling over long periods of time, allowing plants with greater tolerance for soil compaction to thrive.

To ignore ecological processes already present in a region is a well-known phenomenon in shaping cities and the regional landscape. Landscape architect Michael Hough (1990) writes in *Out of Place* that modern cities all tend to look alike despite their regional setting. He argues that the monotony is a reflection of society's indifference to the diversity inherent in ecological systems and in human communities. This is particularly true when one considers the age-old concept of the oasis—the garden. On a small scale, the garden symbolizes the utopian ideal of a fertile and beautiful place in an arid or hostile landscape. As such, it is added to environmental diversity of many places. The urge to expand this ideal to include entire regions, in arid landscape like southern California or Arizona, has been achieved at enormous environmental costs in the withdrawal of groundwater, the transfer of water from distant river systems, and the consequent loss of the essential identity of the natural region (Fig. 2.7).

For example, Los Angeles is situated in a Mediterranean climate with an average annual rainfall of 383 mm. However, nothing is less likely to occur than "average rainfall" in this region. Only 17% of years approach within 25% of the historical average. The average norm turns out to be 7–12-year swings between dry and wet spells. Mike Davis (1998), the author of *Ecology of Fear*, documented research of prehistoric climate that has shown that California has endured two epic megadroughts during the Middle Ages, one of about 220 years and the other about 120 years. The natural vegetation under these climatic conditions is dry grassland, chaparral, and coast scrublands. Most of the Los Angeles coastal plain and interior valleys have been converted to low-density housing, industrial developments, tourist attractions, and a vast agricultural industry, all of which have created an insatiable demand for water. Almost half of domestic and commercial water use in the urban area is for landscape irrigation, which rises to 75% in hotter, lower-density suburbs. All of this irrigation has created an urban

Figure 2.7. Transforming the desert into an oasis near Palm Springs, California. The utopian dream realized at the expense of nature. The contrast between the lush plants imported from high-rainfall regions elsewhere and the native vegetation of the alluvial fan and hills beyond is a powerful expression of unsustainable development and lack of connection to place. Photograph by Robert Landau © 1985, reproduced with permission.

oasis. Looking over the valleys to the surrounding hills, one sees the stark contrast between the two. The hills, browned by summer heat and drought but turning green in spring, represent the sustainable native vegetation that covers this semiarid region. The valleys, perpetually green and lush with lawns, shrubs, and trees, represent the artificial, urbanized landscape sustained by massive irrigation. Most of the vegetation used in urban areas of Los Angeles originates in plant communities in other parts of the world where rainfall is higher.

Reliance on Technology

Why is this so? And why has not more of the natural environment been preserved? In their book, *The California Landscape Garden*, University of California landscape architects Mark Francis and Andreas

Reimann (1999) suggested that reasons first of all lie in our fascination with engineering, which has resulted in vast and elaborate aqueducts, road systems, and so forth. In other words, we do it "because we can" without considering the enormous environmental implications. Readers of landscape architect and planner Ian McHarg's (1969) *Design with Nature* and viewers of his PBS film, "Multiply and Subdue the Earth", will recognize this moral attitude. It derives from the inheritors of the Judaic–Christian tradition who believe that they have received their injunction from Genesis, in which they are given dominion over all life and nonlife and enjoined to subdue the earth.

As Francis and Reimann (1999) wrote, in addition to controlling the environment, technology has allowed us to distance ourselves from the larger world. Many of the materials we use in daily life are from places we ourselves have never been and we know little of who and how they were made. This removes us from a sense of accountability for our actions. This is especially true of water and power, which we take for granted and sometimes squander needlessly.

Furthermore, some of this attitude is encouraged by the media, which tends to separate us from the world around us. Stories in gardening and travel magazines tend to promote contrived landscapes that are simulations of nature—an immaculate lawn in Palm Springs, for example, or an alpine garden in urban San Fernando Valley. These landscapes bear no connection to local and regional ecology or process. The result is a great potential for harm by the overuse of resources required to maintain them and by the displacement of indigenous plant species needed by local wildlife.

Thirty years ago, in *Design with Nature*, McHarg (1969) proposed a system of ecological inventories to help explain the way natural processes may influence regional and urban planning and design. He stated the necessity of understanding natural processes as determinants of land use. His idea is to match geometry and process, inherent in a region's geology, hydrology, soils, and vegetation, with rational land use. In her landmark book, *Waterstained Landscapes*: *Seeing and Shaping Regionally Distinctive Places*, Joan Woodward (2000) provides guidance to promote an understanding of the patterns of processes that shape a region and how to incorporate this understanding in design and planning. In the next chapter, we review those processes of formation.

CHAPTER 3

Regional-Scale Ecosystem Units, Ecoregions

Let us look at the largest regional-scale ecosystem units, or ecoregions. All areas of the earth are made up of distinctive associations of causally related features. Latitude, continental position, and elevation will determine the climate, which, in turn, will affect soil types and vegetation. For example, you will find rainforests and latosolic soils (Oxisols)[4] in tropical wet climates (Fig. 3.1). As the climate changes so do the soil and vegetation in response; and as the climate changes, so do the ecosystems.

The factors that control the climatic effect change with scale. We can distinguish climatic differences and their controls on different levels or scales. For example, we can detect air-temperature differences over a distance of 10,000 km on a global level (related to latitude) and over a few hundred meters in mountain areas (related to exposure or aspect).

At the macroscale, ecoregions (macroecosystems) correspond to the large climatic regions where climatic conditions are relatively uniform. These regions are delineated by ignoring the smaller-scale effects of landform and vegetation. Locating boundaries of broad-scale ecosystems requires taking into account visible and tangible expressions of climate such as vegetation. Generally, each climatic region is associated with a single plant formation (such as savanna) and is characterized by a broad uniformity both in appearance and in composition of the dominant plant species.

At smaller scales, surface features break up the climatic regions into local climates and their associated ecosystems. Ecoregions are subdivided into areas we call **landscape mosaics** (Fig. 3.2). These, in turn, are subdivided into small-scale units known as **sites**. The smallest, sites, are local ecosystems (microecosystems) commonly recognized by foresters and range scientists. They are the size of hectares. Linked sites

Tropical wet climates

Tropical rainforests

Latosolic soils

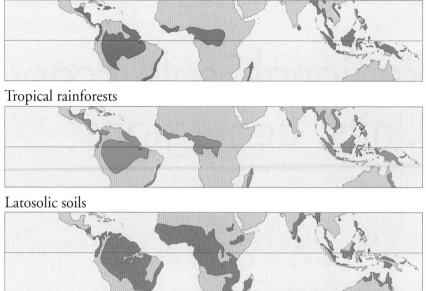

Figure 3.1. Spatial correspondence in the tropics among broad categories of climate, vegetation, and soils. Climate from Trewartha; vegetation after Eyre, Küchler, and others; soils based on numerous sources, including Soil Conservation Service. Redrawn from *Physical Elements of Geography*, 5th ed., Frontispiece, Plate 5, Plate 6. Copyright © 1967 by McGraw-Hill Companies; reproduced with permission.

create a landscape mosaic (mesoecosystem) that looks like a patchwork when seen from above. A landscape mosaic is made up of spatially contiguous sites distinguished by material and energy exchange. They range in size from 10 to several thousand kilometers.

A mountain range is a classic example of a landscape mosaic. A lively exchange of materials occurs among the component ecosystems of a mountain range: Water and products of erosion move down the mountains; updrafts carry them upward; animals can move from one ecosystem to the next, and seeds are easily scattered by the wind or distributed by birds.

Ecoregions Versus Other Land Divisions

In contrast to other categories of land division, such as **physiographic regions** or **biotic areas** (also called biotic provinces and bioregions), an ecoregion is based on both living and abiotic features. A geologist

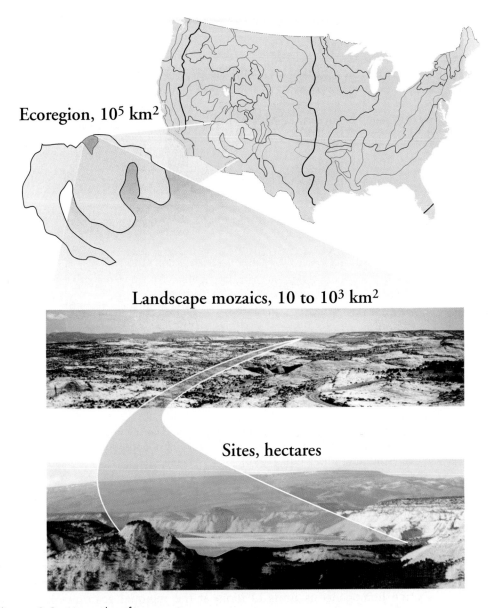

Ecoregion, 10⁵ km²

Landscape mozaics, 10 to 10³ km²

Sites, hectares

Figure 3.2. Hierarchy of ecosystems.

might look at a given area in terms only of geologic formations and structures (Fig. 3.3, top). In fact, a geologist produced one of the best known physiographic maps of the United States with this perspective (Fenneman 1928). Where major physiograpic discontinuities occur, where mountains meet plains, or where **igneous rock** change to sedi-

Figure 3.3. Landscape and different types of diversity in landscape ecosystems: geodiversity of a landscape (*top*), biodiversity of a landscape (*bottom*). From Concepts by Leser and Nagel (1998).

mentary strata, the boundaries of these units often coincide with changes in the vegetation and associated fauna, or **biota**. In areas of little relief, such as the Great Plains, there tends to be little or no correlation between the geologist's concept of physiography and ecology.

Likewise, a biologist (cf. Dice, 1943) might look at the same area in terms of the spatial pattern of the biota (Fig. 3.3, bottom). Large, relative homogeneous units of biota at the regional scale are known as **biomes** (Clements and Shelford 1939). Heinrich Walter (1985) refers to these as **zonobiomes**, because they are based on large climatic zones. Subdivisions of biomes have been mapped by Miklos Udvardy (1975) after work started by Raymond Dasmann (1973) and are called **biogeographic provinces**. However, biota is constantly changing due to disturbance and **succession**. For example, fires or timber harvesting may destroy a forest, causing fauna dependent on the forest to migrate. As the process of succession restores the forest to predisturbance conditions, the fauna will repopulate the forest. Moreover, the geographic distribution of animal species or communities may change due to hunting, independent of habitat loss.

We need to base ecosystem boundaries on the factors that control ecosystem distribution at various scales, rather than on present biota in order to screen out the effects of disturbance and natural succession. In this way, ecosystems can be recognized, compared, and worked with regardless of the present land use or other disturbance. The potential of any system makes it possible to understand and manage it wisely. One way to get at this potential is through the concept of the **climax**. This concept, developed largely by Frederick Clements (1916), recognized that vegetation develops through a series of stages until the whole region is clothed with a uniform plant and animal community. The final stage is determined solely by the climate and is known as a **climatic climax vegetation**. Although theoretically possible, because of variations in local environments, particularly in soil parent material, such a region is more likely to have several or many climax vegetation types. This realization has led to the polyclimax concept (Tansley 1935), which recognizes that the climax vegetation of a region consists of not just one type but a mosaic of climaxes controlled primarily by soil conditions. These are the **edaphic climaxes**.

Thus, if we are going to look at that area in terms of the ecological consequences of human activities, we must understand the full range of features of both the geologists and biologists and more (Fig. 3.4). We

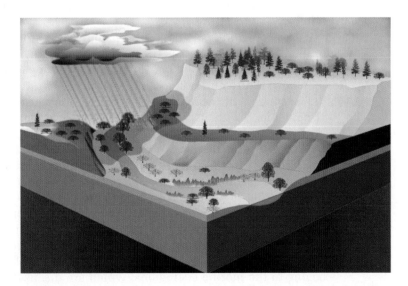

Figure 3.4. Ecological diversity of a landscape. From concepts by Leser and Nagel (1998).

must also look at it from the standpoint of an ecosystem—not just locally, but also including the surrounding area.

With regard to their delimitation and the purpose behind it, namely to create a system for an ecological division of the world, the term "ecoregion" is comparable to those regions referred to by other authors as "major natural regions" (Herbertson 1905), "landscape belts" (Passarge 1929), "landscape zone" (Isachenko 1973), "ecozone" (Schultz 1995), "morphoclimatic zones" (Tricart and Cailleux 1972), and so forth. The former differs from the latter in that ecoregions are based explicitly on the distribution of ecosystems. However, the concept of ecoregions is much older. The ancient Greeks recognized such a concept. In the eighteenth century, Baron Alexander von Humbolt provided an outline of latitudinal zonality and high-altitude zonality of the plant and animal world in relation to climate (Fig. 3.5). The significant work of Dokuchaev (1899) developed the theory of integrated concepts. He pointed out that within the limits of extensive areas (zones), natural conditions are characterized by many features in common, which change markedly in passing from one zone to another. One could consider the efforts of C. Hart Merriam (1898) to define life zones of the United States as one approach to delineating regions. He described seven transcontinental belts, or life zones, based on associations of plants and animals. His work established that these natural zones were suitable to certain varieties of crops. Later, the United States Department of Agriculture (U.S. National Arboretum 1965) developed the Plant Hardiness Zone Map, which divides the country into 11 zones and shows whether a crop or garden plant will survive the average winter.

Each ecoregion is characterized usually by a single climax, but two or more climaxes may sometimes be represented within a single ecoregion. This often happens on mountains, where each altitudinal zone may have a different climax.

The concept of "ecoregion" differs from that of "biome," for a biome is coincident with its climaxes. Every area having the same climax, however far detached from the main area of that climax, seems to belong to the same biome. An ecoregion, on the contrary, is never discontinuous (except on marine islands), although ecological communities somewhat similar to those characteristic of the particular ecoregion may exist far beyond its boundaries in other parts of the world.

Each ecoregion comprises both the climax communities and all of the successional stages within its geographic area, and it thus includes the freshwater communities. It does not, however, include the marine communities that may lie adjacent to its shores. These communities belong to the marine ecoregions, which are discussed elsewhere (Bailey 1998a).

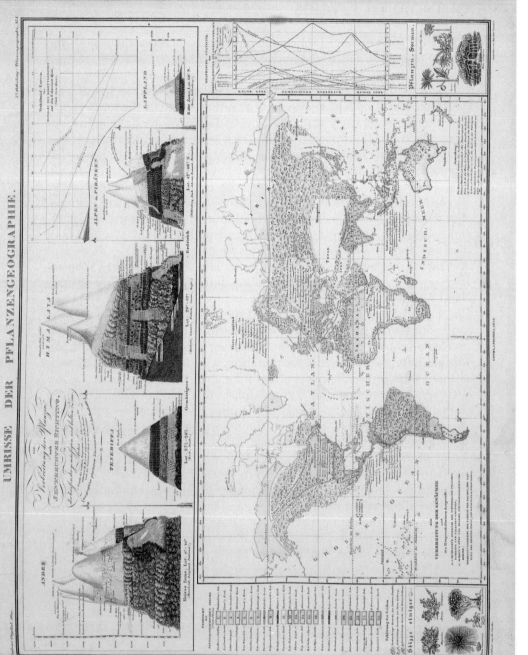

Figure 3.5. Map of ecological geography, derived from the work of German geographer Alexander von Humboldt and Danish botanist Joakim Frederick Schouw, is from Heinrich Berghaus's three-volume *Physikalischer Atlas* (1845). From the atlas collection of the Library of Congress; reproduced with permission.

39

Analysis of Natural Cycles

Climatic, **geomorphic**, biotic, and cultural processes determine the ecoregions. Each has a distinctive cycle of natural processes, including life, climate, hydrology, erosion, deposition, and, even, fire.

Each region has a different **climatic regime**, defined as the diurnal and seasonal fluxes of energy and moisture. We can illustrate different climatic regimes by studying climate diagrams, or climographs. For example, tropical rainforest climates lack seasonal periodicity, whereas mid-latitude **steppe** climates have pronounced seasons (see the diagrams for Singapore and Colorado Springs, Colorado, in Appendix B). As the climatic regime changes, so does the **hydrologic cycle**, as reflected in the streamflow of rivers located in different climatic regions. For example, no water flows in creeks located in the warm, dry, summer region of California during summer and fall, but in winter and early spring, groundwater contributes to streamflow.

Climate profoundly affects landforms and erosion cycles. Such effects are remarkably evident when we contrast the angularity of aridland topography of the Colorado Plateau with the rounded slopes of the humid Blue Ridge Mountains of the eastern United States. Plants and animals have adjusted their life patterns to the basic environmental cycles produced by the climate. Whenever a marked annual variation occurs in temperature and precipitation, a corresponding annual variation occurs in the life cycle of the flora and fauna. Annual cycles are very apparent in the tropical grasslands and in mid-latitudes. The rainforest and polar deserts are about the only regions that do not experience annual changes. Many plants and animals adjust to the changing length of day throughout the year. Reproduction, dormant periods, color changes, migration, and many other life patterns are adapted to moisture cycles (Fig. 3.6). In the past, forest fires occurred at different magnitudes and frequencies in different climatic–vegetation types. In the **boreal forest**, for example, infrequent large-magnitude fires carried the flames in the canopy of the vegetation, killing most of the forest. Other regions, such as the lower-elevation ponderosa pine forest in the western United States, had a regime of frequent, small-magnitude, surface fires. Here, the burning was restricted to the forest floor and most mature trees survived.

Process of Differentiation

To delineate these regions and to understand how and why they are distributed, we must understand the processes of how they are differentiated. This is important in understanding their dynamics and how

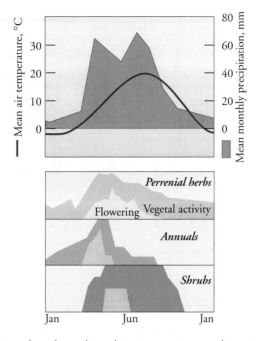

Figure 3.6. Annual cycles adapted to temperature and moisture in a climax steppe community. Top figure is redrawn from Daubenmire (1968); bottom figure is redrawn from Walter et al. (1975). Fig. 4, p. 11 from *Plant Communities* by Rexford Daubenmire. Copyright © 1968 by Rexford Daubenmire. Reprinted by permission of Pearson Education, Inc.

they respond to management. These processes are presented in my previous books (Bailey 1996, 1998a), but let us review a bit.

Two primary sources of energy and their resultant processes differentiate the Earth's surface into ecoregions: one is external, provided by the sun (Fig. 3.7). The sun's energy interacting with our atmosphere creates climates. The factors controlling spatial variation in climate (and therefore ecosystems) occur at several levels. At the global level, ecosystem patterns are controlled by the **macroclimate** (i.e., the climate that lies just beyond the local modifying irregularities of landform and vegetation) and are related to the variation in solar radiation with latitude. The low latitudes or tropics receive more solar radiation than do middle and higher latitudes.[5] Only about 40% as much solar energy is received above the poles as above the equator. If the Earth were of uniform composition (either land or water), there would be simple, east–west zones of climate resulting from variation in the amount of solar radiation that reaches different latitudes (Fig. 3.8). They would owe their differentiation to the varied effects of the sun, instead of the character of the surface.

Figure 3.7. External (solar) energy source.

However, the Earth's surface is rather heterogeneous, divided into large masses of land (the continents), water (the oceans), and ice (the polar regions). At the continental level, differential heating between land and sea gives rise to distinctive continental climates with wider ranges of temperature, lower humidity, and more variable precipitation than marine climates. At any given latitude, the summers are warmer and the winters colder on land than on the oceans. The warmest months increase with increasing latitude and distance from the ocean. This forms a distinction between marine and continental climates (Fig. 3.9).

The other primary energy source is the heat generated within the Earth itself (Fig. 3.10). It drives mantle convection and produces plate tectonics, causing mountain building. Mountainous areas are associ-

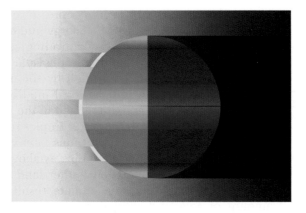

Figure 3.8. Latitudinal climatic zones that would result if the Earth were simply a granite sphere with an atmosphere.

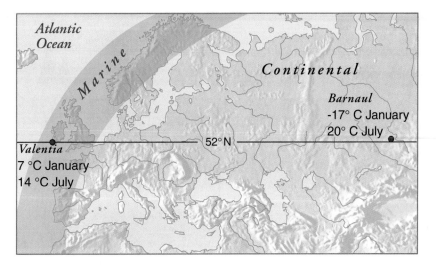

Figure 3.9. Marine and continental climates. Redrawn from *Physical Geography in Diagrams,* by R.B. Bunnett. Copyright © 1968 by R.B. Bunnett. Reprinted by permission of Pearson Education Limited.

ated with the margins of **crustal plates**, and the great elevations result from the upwarping of the crust along the plate boundaries and the upwelling of **magma** that forms the volcanic peaks and massive lava flows. These mountains are arranged without conforming at all to the orderly latitudinal zones of climate. They cut irregularly across latitudinally oriented climatic zones. For example, we find mountains in the cold deserts of Antarctica as well as near the equator.

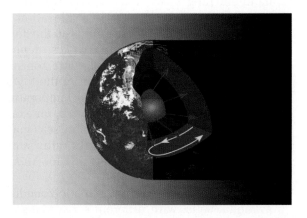

Figure 3.10. The Earth's internal energy sources. Base from Mountain High Maps®. Copyright © 1997 by Digital Wisdom, Inc.

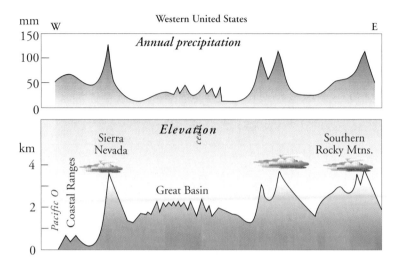

Figure 3.11. Effect of elevation on precipitation across the western United States at approximately 38° north. Redrawn from Bailey (1941).

The resulting mountains on the continents modify the climatic pattern that would otherwise develop on a flat continent. Mountain climates are vertically differentiated, based on the effects of changes in elevation. Air cools while ascending the mountain slope, and its capacity to hold water decreases, causing an increase in rain and snow (Fig. 3.11). The thin, dry air loses heat rapidly as it ascends, and after sunset, temperatures plummet.

The arrangement of the ecological climate zones depends largely on latitude and continental position. This pattern, however, is overlain by mountain ranges, which cut across latitudinally oriented climatic zones to create their own ecosystems. Elevation creates characteristic ecological zones that are variations of the lowland climate. Mountains show typical climatic characteristics, depending on their location in the overall pattern of global climatic zones. In other words, elevation produces a predictable variation of the lowland climate, especially in a climatic regime (seasonality of temperature and precipitation). The mountain ranges of Central America, for example, experience the same year-round, high-energy input and seasonal moisture regime consisting of a relatively dry winter and a rainy summer typical of their neighboring lowlands (see the diagrams for Mexico City and San Salvador in Appendix B).

Every mountain within a climatic zone has a typical sequence of elevational belts, with different ecosystems at successive levels: generally montane, alpine, and **nival**, but exhibiting considerable differences according to the zone where they occur. When a mountain extends

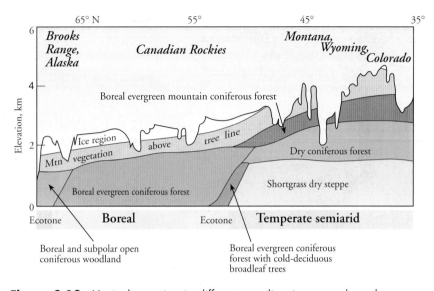

Figure 3.12. Vertical zonation in different ecoclimatic zones along the eastern slopes of the Rocky Mountains. Redrawn from Schmithusen (1976).

over two or more climatic zones, it produces different vertical zonation patterns (see Fig. 3.12, which compares locations in the Rocky Mountains). In the temperate semiarid climatic portion, the lowermost zone is a short-grass dry steppe basal plain; this is followed by a montane zone of ponderosa pine, Douglas fir, spruce, and fir. Above is the subalpine zone, followed by alpine tundra, and then perennial ice and snow. This sequence of elevational zones repeats on mountain ranges throughout the lowland, semiarid climatic zone.

Between the individual elevational belts, a lively exchange of materials occurs: Water and the products of erosion move down the mountains; updrafts and downdrafts carry dust and organic matter; animals move easily from one belt into the next; and wind and birds spread seeds. The belts, as a result, are interconnected and the geographic area over which a sequence of belts extends is considered to be a large ecological unit, an ecoregion. In this sense, we do not treat the montane forest belt as a separate ecoclimatic zone. The montane belt is only one member of the total sequence of elevational belts. Montane belts in mountainous areas of different climatic zones are just as distinct from one another as the montane belt is from other elevational belts in the same zone.

The combined effects of latitude, continental position, and mountains form the world's regional-scale ecosystems, or ecoregions (Fig. 3.13). Each ecoregion includes areas in different parts of the world that are broadly similar in climate, surface features, and vegetation. For ex-

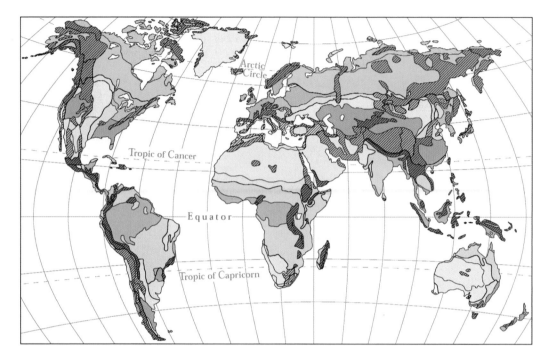

Figure. 3.13. A scaled-down, simplified version of the ecoregions map of the continents, as mapped by Bailey (1989).

ample, the tropical/subtropical steppe division of the dry domain is found on all continents (Fig. 3.14). These steppes typically are grass-lands of short grasses and herbs with local shrub and woodland. Pinyon-juniper woodland, for example, grows on the Colorado Plateau of the United States. The soils are Mollisols or Aridisols.

This approach is based on understanding the formative processes that operate to differentiate ecoregions at various scales in a hierarchy. The units derived from such an approach are termed *genetic*. As Rowe (1979) pointed out, the key to the placing of map boundaries on eco-logical maps is the understanding of genetic processes. We can only comprehend a landscape ecosystem if we know how it originated or evolved. That is why Huggett (1995) suggested that the approach is *evolutionary* as well. This approach is distinguished from another ap-proach that attempts to create ecoregion frameworks that would be more objective and repeatable. The frameworks are based on the over-lay of ecosystem component maps (e.g., climate, soil, etc.) using a GIS overlay operation. Lines are shifted to coincide when the boundaries are divergent. The maps, however, may be so inaccurate or unable to capture significant units of productivity or ecological response that they could lead to imperfect or false conclusions (Bailey 1988b). Other

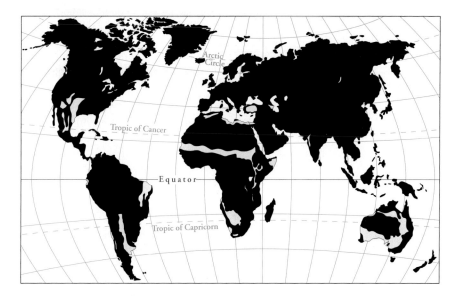

Figure 3.14. Global pattern of the tropical/subtropical steppes.

researchers have attempted to develop quantitative classifications of ecoregion units as well. These involve using multivariate clustering of grid cells or sample points. A map is produced by drawing lines around cells or points of similar class. However, as Rowe (1980) pointed out, the units derived from such a process are not necessarily ecological. Ecological units can be comprehended only as wholes that have some process significance. For example, a floodplain is a pattern of spatially associated, but *unlike* land units (cells). The floodplain consists of the active channel, abandoned channel, islands, lakes, wetlands, levees, and so forth. Each unit has different characteristics, but is united with the others by common processes of development, namely cyclic inundation, erosion, meandering, and deposition.

Ecoregional Mapping

The map Ecoregions of the Continents (Bailey 1989) was produced using a worldwide classification developed (Bailey 1983, 1989, 1998a; Bailey and Hogg 1986) from concepts advanced by John Crowley (1967). The general principle followed has been to identify ecosystem regions of continental scale based on macroclimate. Macroclimates are among the most significant factors affecting the distribution of life on Earth. As the macroclimate changes, the other components of the ecosystem change in response. Macroclimates influence soil formation

Table 3.1. Köppen's climatic classification

- Based on
  - Monthly and annual temperature and rainfall
  - Thermal and moisture limits for vegetation
- Five main groups
  - Four thermally defined
  - *Dry* is not
- Subdivided into 11 types based on
  - Seasonality of precipitation
  - Degree of dryness or cold

and help shape the surface topography, as well as affecting the suitability of a given system for human habitation. As a result, ecosystems of different macroclimates differ significantly.

Based on macroclimatic conditions and on the prevailing plant formations determined by those conditions, I subdivided the continents into ecoregions with three levels of detail. Of these, the broadest, *domains*, and within them *divisions*, are based largely on the broad ecological climate zones of Wladimir Köppen (as modified by Trewartha 1968, Table 3.1, Appendix A).[6] Thermal and moisture limits for plant growth determine their boundaries. There are four groups. Three are humid, thermally differentiated: polar, with no warm season; humid temperate, rainy with mild to severe winters; humid tropical, rainy with no winters. The fourth, dry, is defined on the basis of moisture alone and transects the otherwise humid domains. Within these groups are 15 types of climate based on seasonality of precipitation or on degree of dryness or cold; for example, within the humid tropical domain, rainforests with year-round precipitation can be distinguished from savannas with winter drought. Divisions correspond to these types. Each division is clearly defined by a particular type of climate diagram that helps explain the conditions that create them (see Appendix B for stations thought to be representative of each division). Table 3.2 lists climate, vegetation, and soils types associated with each division. For more information, including illustrated, detailed descriptions of the divisions, see my related book (Bailey 1998a).

The climate is not completely uniform within divisions, so that a further subdivision can be undertaken. Within the dry climates, for example, there is a wide range of degree of aridity, ranging from very dry deserts through transitional levels of aridity in the direction of adjacent moist climates. We refer to these as **climate subtypes**. The subtypes largely correspond to major plant formations (e.g., broad-leaved forest), which are delimited on the basis of macro features of the vegetation by concentrating on the life-form of the plants. They form the basis for subdividing ecoregion divisions into *provinces* and are based

Table 3.2. General environmental conditions for ecoregion divisions

Name of division	Equivalent Köppen–Trewartha climates	Zonal vegetation	Zonal soil type[a]
110 Icecap			
120 Tundra	Ft	Ice and stony deserts: tundras	Tundra humus soils with soilifluction (Entisols, Inceptisols, and associated Histosols)
130 Subarctic	E	Forest–tundras and open woodlands; tayga	Podzolic (Spodosols and Histosols)
210 Warm Continental	Dcb	Mixed deciduous–coniferous forests	Gray-brown Podzolic (Alfisols)
220 Hot Continental	Dca	Broad-leaved forests	Gray-brown Podzolic (Alfisols)
230 Subtropical	Cf	Broad-leaved coniferous evergreen forests; coniferous broad-leaved semievergreen forests	Red-yellow Podzolic (Ultisols)
240 Marine	Do	Mixed forests	Brown forest and gray-brown Podzolic (Alfisols)
250 Prairie	Cf, Dca, Dcb	Forest–steppes and prairies; savannas	Prairie soils, Chernozems (Mollisols)
260 Mediterranean	Cs	Dry steppe; hard-leaved evergreen forests, open woodlands and shrub	Soils typical of semiarid climates associated with grasslands
310 Tropical/ subtropical steppe	BSh	Open woodland and semideserts; steppes	Chestnut, brown soils, and Sierozems (Mollisols, Aridisols)
320 Tropical/ subtropical desert	BWh	Semideserts; deserts	Desert (Aridisols)
330 Temperate steppe	BSk	Steppes; dry steppes	Same as BSh
340 Temperate desert	BWk	Semideserts and deserts	Same as BWh
410 Savanna	Aw, Am	Open woodlands, shrubs and savannas; semi-evergreen forest	Latosols (Oxisols)
420 Rainforest	Ar	Evergreen tropical rain forest (selva)	Latosols (Oxisols)

[a]Great soil group. Names in parentheses are Soil Taxonomy soil orders (USDA Conservation Service 1975); described in the Glossary.

on a number of sources, including a world map of landscape types (Milanova and Kushlin 1993).

With few exceptions, the climate subtypes largely correspond to **zonal** soil types and zonal vegetation. For example, boreal coniferous forests (tayga) and podzolic (Spodosols) soils correspond to the subarctic climate in the Köppen–Trewartha system. Zonal soil types and vegetation occur on sites supporting climatic climax vegetation. Such sites are uplands (i.e., sites with well-drained surface, moderate-surface slope, and well-developed soils).

Mountains exhibiting **elevational zonation** and the climatic regime of the adjacent lowlands are distinguished according to the character of the zonation by listing the altitudinal zones present. Such mountainous environments are termed *mountain provinces.* More details about mapping procedures are presented elsewhere (Bailey 1983).

Pattern within Regions

Within the same macroclimate, broad-scale landforms break up the east–west climatic pattern that would occur otherwise and provide a basis for further differentiation of ecosystems—the landscape mosaics mentioned earlier. The character of a landscape mosaic with identical geology will vary by the climate zone. For example, vertical limestone would form quite different landscapes in a subarctic climate than in hot and arid climates. Limestone in a subarctic climate occurs in depressions and shows intense **karstification**, whereas in hot and arid climates, it occurs in marked relief with a few cave tunnels and canyons inherited from colder **Pleistocene** time (Fig. 3.15).

Landforms (with their geologic substrate, surface shape, and relief) influence place-to-place variation in ecological factors such as water availability and exposure to radiant solar energy. Through varying height and degree of inclination of the ground surface, landforms interact with climate and directly influence hydrologic and soil-forming processes.

In short, the best correlate of vegetation and soil patterns at mesoscales and microscales is landform, because it controls the in-

Hot and arid Arizona-Sonoran deserts, U.S.   *Humid subarctic Norwegian Lapland*

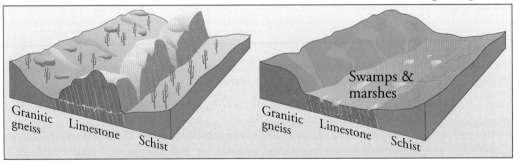

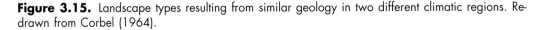

Figure 3.15. Landscape types resulting from similar geology in two different climatic regions. Redrawn from Corbel (1964).

tensities of key factors important to plants and to the soils that develop with them (Hack and Goodlet 1960; Swanson et al. 1988). The importance of landform is apparent in a number of approaches to the classification of forest land (e.g., Barnes et al., 1982). Even in areas of relatively little topographic relief, such as the glacial landforms of the upper Midwest of the United States, landform explains a great deal of the variability of ecosystems across the landscape (Host et al. 1987).

Landforms come in all scales and in a great array of shapes. On a continental level within the same macroclimate, several broad-scale landform patterns commonly break up the zonal pattern. They range from nearly flat plains to rolling, irregular plains, to hills, to low mountains, to high mountains (Hammond 1954). They are identified on the basis of three major characteristics: (1) relative amount of gently sloping (<8%) land, (2) local relief, and (3) generalized profile (i.e., where and how much of the gently sloping land is located in valley bottoms or in uplands).

According to its physiographic nature, a landform unit consists of a certain set of ecosystem sites. A delta has differing types of ecosystems from those of a moraine landscape next to it. Within a landscape mosaic, the sites are arranged in a specific pattern. The tablelands of the west-central part of the North American continent are a case in point. For example, the Colorado Plateau is made up of various site-specific ecosystems, including valleys of various sizes, smooth uplands, stream channels (mostly dry), individual slopes, terraces, sandbars in the stream channels, and several small and shallow depressions in the uplands.

Although the distribution of ecoregions is controlled by macroclimate and broad-scale landform patterns, local differences are controlled chiefly by microclimate and ground conditions, especially moisture availability. The latter is the edaphic (related to soil) factor.

Within a landform, slight differences in slope and aspect modify the macroclimate to **topoclimates** (Thornthwaite 1954). As outlined by Hills (1952), three classes of topoclimate are commonly recognized: normal, hotter than normal, and colder than normal (Fig. 3.16). We refer to the ecosystems controlled, and partially defined, by topoclimate as *site classes*, following Hills (1952).

In differentiating local sites within topoclimates, soil moisture regimes provide the most significant segregation of the plant community. A sequence of moisture regimes, ranging from the top to the bottom of a slope, is as follows: drier, moist (normal), and wetter (Fig. 3.17). It may be referred to as a soil catena, or **toposequence** (Major 1951). Exposure to wind also influences soil moisture. The existence of small relief forms substantially affects the movement of air masses; it changes the direction and velocity of winds near the ground, thus contributing to the redistribution of rainfall. The windward hill slopes

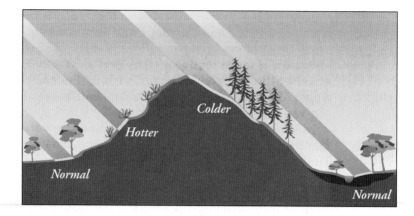

Figure 3.16. Slope and aspect affect temperature, creating topoclimates.

usually receive less rain than the lee slopes. From the hilltops, snow blows into depressions, where it accumulates and remains 1–2 weeks longer than on elevated sections.

The **lithology** of the bedrock also affects vegetation patterns. Different kinds of rock vary in their resistance to erosion, their hydrologic properties (porosity, permeability, and so on), and chemistry. This affects not only the topography, but also soil formation and subsequent moisture content. This is particularly well illustrated in the semiarid regions with sedimentary rock, such as the Colorado Plateau. Here, the bedrock is interbedded sandstone and shale. The shale erodes more easily, forming soils with higher moisture. Such soils support a more dense vegetation consisting of lightly but scattered grasses, shrubs, and

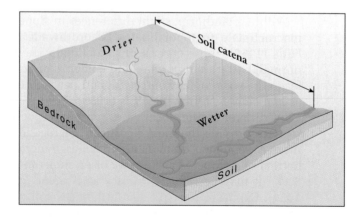

Figure 3.17. Variation in moisture creates a toposequence or catena of soil moisture regimes.

small trees. This banding, a **lithosequence**, is caused by the preference of vegetation for greater moisture of slopes underlain by rock with slightly greater moisture.

Deviations from normal topoclimate and mesic soil moisture occur in various combinations within a region and are referred to as *site types* (Hills 1952). As a result, every regional system—regardless of size or rank—is characterized by the association of three types of local ecosystem or site type: zonal, azonal, and intrazonal.

Zonal site types. These sites are characterized by a normal topoclimate and moist soil. They are typically located on the well-drained uplands in the landscape.

Azonal site types. These sites are zonal in a neighboring zone but are confined to an extrazonal environment in a given zone. For instance, in the Northern Hemisphere, south-facing slopes receive more solar radiation than north-facing slopes; thus, south-facing slopes tend to be warmer, drier, less thickly vegetated, and covered by thinner soils than north-facing slopes. In arid mountains, the south-facing slopes are commonly covered by grass, whereas steeper north-facing slopes are forested. Azonal sites are hotter, colder, wetter, or drier than zonal sites.

The size, direction, and configuration of valleys and basins are also important in determining azonal conditions. Valleys, for example, produce their own wind systems. At night, air over the valley slopes becomes its coldest and heaviest and thus is carried by gravity to the bottom of the valley. Vegetation refects these air movements. Cold-air drainage (the cold downdrafts) in the montane zone of the Rocky Mountains creates grassy areas in the valleys that are too cold for tree growth. Early settlers referred to these treeless areas as "parks". These air movements also affect agricultural ventures. Grassy areas in citrus groves in Florida are slightly lower than the surrounding terrain. It can be 2°C colder than the higher ground— just cold enough to kill the trees that were planted there.

Intrazonal site types. These sites occur in exceptional situations within a zone. They are represented by small areas with extreme types of soil and intrazonal vegetation. Soil influences vegetation to a greater extent than climate; thus, the same vegetation forms may occur on similar soil in a number of zones. We can differentiate them into five groups:

1. *Unbalanced chemically* is the first site type. Some examples from the United States are the specialized plant stands on serpentine (magnesium-rich) soils in the California Coast Ranges. Other examples are the belts of grassland on the lime-rich black

belts of Alabama, Mississippi, and Texas, and the low mat salt-bush on shale deserts of the Utah desert, which contrasts with upright shrubs on adjacent sandy ground. The kind and amount of dissolved matter in groundwater also affects plant distribution. This is especially obvious along the coasts and along edges of desert basins, where the water is brackish or saline. Plants adapted to moist saline ground are called halophytes.

2. *Very wet* sites occur where the groundwater table controls intrazonal plant distributions. The plants of these sites are phreatophytes, plants that send roots to the water table. Examples include riparian zones in the deserts of the southwestern United States, such as a cottonwood floodplain forest and the cypress and tupelo forests of the Southeast.

3. *Very dry* sites with sandy soils, because of limited moisture-holding capacity, are drier than the general climate. At the extreme, sand dunes fail to support any vegetation because they are too dry. Good examples include dunes along the coasts of oceans and lakes and the sand dunes that have formed in arid interior regions of the continents.

4. *Very shallow* sites are another type. Soil depth, as a factor in plant distribution, may be controlled by depth to a water table or depth to bedrock. Vegetation growing along a stream or pond differs from that growing some distance away, where the depth to the water table is greater. Examples of the influence of depth to bedrock on plant distribution can be seen in mountainous areas where bare rock surfaces that support only **lichens** are surrounded by distinctive flowering plants growing where thin soil overlaps the rock and is, in turn, surrounded by forest where the soil deepens.

5. *Very unstable* sites are areas where gravity combined with high relief, steep slopes, weak bedrock, excessive groundwater, earthquake shocks, and undercutting causes landslides. These slides include slump-earthflows, rockslides, rockfalls, mantle slides, and mudflows. Commonly, these slides produce anomalies in the topography and vegetation.

Figure 3.18, in a simplified way, illustrates how topography, even in areas of uniform macroclimate, leads to differences in local climates and soil conditions. This example is from southern Ontario, Canada (Hills 1952). On level or moderately rolling areas where the soil is well drained but moist, a maple–beech community (sugar maple and beech being the dominant plants) is the terminal succession. Because we find this type of community repeatedly in regions wherever land configu-

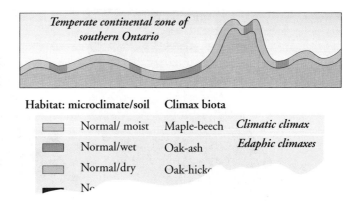

Figure 3.18. Forest climaxes relate to topography in the temperate continental zone of southern Ontario, Canada. (Diagram is truncated, showing only three of nine possible climaxes.) Simplified from Hills (1952). In Odum (1971). From *Fundamentals of Ecology*, 3rd edition, by E.P. Odum, Copyright © 1971. Reprinted with permission of Brooks/Cole, an imprint of the Wadsworth Group, A Division of Thomson Learning. Fax 800 730-2215.

ration and drainage are moderate, the maple–beech community is judged to be the normal, unmodified climax of the region. Where the soil remains wetter or drier than normal, a somewhat different end community occurs, as indicated. The climatic climax theoretically would occur over the entire region except for topography leading to different local climates, which partially determines edaphic conditions. On these areas, different edaphic climaxes occur; climatic climaxes occur only on mesic soils.

Slopes of similar physical characteristics will be found in various ecoregions and will support different ecosystems because of the different climates. For example, a certain slope in the Arctic will support low-growing shrubs and **forbs**, whereas an identical slope in a warm continental ecoregion will have dense broad-leaf and evergreen forests (Fig. 3.19). On the other hand, the same vegetation may exist in different ecoregions due to **compensation factors**, such as soil, that override the climatic effect. For example, in the High Plains and southwestern United States, forests extend into arid and semiarid regions along streams because of the extra water supply (Figure 3.20).

In Ontario, Canada, Hills (1960) and his co-workers (see Burger 1976) have defined regions (called "site regions") within which specific plant successions occur upon specific landform positions. Conversely, similar landforms (relief and geological materials) within different regions will support different plant successions. The different vegetation/landform relationships in various site regions are a reflection of dif-

Arctic   Warm continental

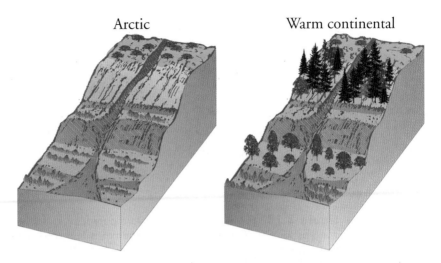

Figure 3.19. Examples of ecosystem diversity in different ecoregions. Adapted, in part, from Leser and Nagel (1998).

Figure 3.20. Riparian forests near Scottsbluff, Nebraska. Author's photograph.

Table 3.3. Change of preferred positions of three tree species[a] on the normal moisture regime, but with three different topoclimates for the seven most easterly site regions of Ontario, Canada[b]

| Site region | Hotter | Topoclimate | |
		Normal	Colder
1	P		
2	P		P
3	P		P
4	A	P	P
5	A		A P
6	C		A P
7			C A

[a]P = *Pica glauca* (white spruce); A = *Acer saccharum* (sugar maple); C = *Carya ovata* (shagbark hickory).
[b]From Burger (1986)

ferences in regional climate. Table 3.3 demonstrates how, for the same normal soil moisture condition but with three different topoclimates (i.e., normal, hotter than normal, and colder than normal), three species of trees (white spruce, sugar maple, and hickory) change their preferred positions in the seven most eastern site regions in Ontario. With these changes, related changes also occur in the vigor of other tree species, ecosystem productivity, and type of ground vegetation, which competes with forest regeneration.

Soil–site relationships have been extensively studied in the southern Appalachian Mountains and the relationship between productivity and soil and topographic variable has been quantified for many important tree species. McNab (1990) found that the suitability of sites in this region for certain forest types can be predicted accurately just on the basis of topographic data measured on site. Suitability of sites can be predicted by means of geographic information systems, because values of the topographic variables can be calculated from a digital elevation database.

In summary, we can interpret the patterns of continental ecosystems through the primary factor of climate. Regional ecosystems, or ecoregions, are areas of homogeneous macroclimate. The arrangement of these ecoregions is regular and predictable because the controlling factors are the same for each. We can predict the kind of system that will be found in any particular place on Earth if we know the latitude, relative continental position, and elevation. Likewise, within each ecoregion is a characteristic pattern of sites that recur in a predictable way, as a result of the nature of the soil and surface. We discuss the role of ecoregions in sustainability in the next chapter.

An Ecoregional Approach to Sustaining Ecosystems

Ecology-Based Design

Because ecology-based design responds to the ecoregion, we must consider the relationships among soils, vegetation, materials, culture, climate, and topography in a particular region. In other words, the ecoregional setting needs to be taken into account in our designs.

We can begin by looking for information about successful ecological designs in other areas around the globe with the same kind of ecoregion. In their book, *Ecological Design*, Van der Ryn and Cowan (1996) referred to this approach as bioregional-level planning. They described how, in Austin, Texas, Pliny Fisk and the Center for Potential Building Systems (Max's Pot) have rigorously pursued this approach (Fisk 1983). The most fully realized project of Max's Pot to date is the Laredo Demonstration Blueprint farm, a 2-acre farm at the edge of Laredo, Texas. This project responds to the climate, mineral resources, vegetation, and soil of its region, which lies in the tropical/subtropical steppes between the arid deserts of the Southwest and the subhumid **prairie** grasslands.

Max's Pot begins every project by looking at the ecologically appropriate designs indigenous to other ecological regions around the globe that have similar climate and vegetation (tropical/subtropical steppes shown in Fig. 3.14). In Texas, the scrubby mesquite tree is regarded as a nuisance and ruthlessly cleared away. In the badlands of Argentina, though, it has long been used for floor tiles. On the Laredo farm, mesquite tiles are used for paving. In a similar fashion, the design of the farm's cooling towers in the storage sheds were borrowed from Iran.

The design of the Laredo farm clearly grows from the character of its place, responding to the area's unique set of factors: generators utilize the wind, **cistern** catchment systems capture rainfall, the crop-shading system and cooling towers provide protection from the sun, and agricultural wastes are treated before reaching the Rio Grande River. It also uses local resources: vegetation (mesquite tiles), minerals (e.g., **caliche**), and locally produced wastes (straw bales). By responding in an information-rich, energy-poor, and materials-frugal manner to a demanding landscape, the Laredo farm minimizes destructive ecological impacts.

Fisk and his partner, Gail Vittori, also have applied this approach to sustainable living in their "ecodynamic" home and office in Austin, which was recently profiled in *Natural Home* magazine (Lawrence 2000). The building's walls are made of adobe, rammed earth, straw from nearby fields, and Calcrete, a mixture of indigenous caliche and concrete containing fly ash from nearby coal-burning facilities. Recycled materials frame the doors and windows. Solar panels on the roof provide electricity; rainwater is harvested in cisterns for drinking and bathing. Composting toilets treat wastewater, and graywater is funneled into a wetland garden, which includes cattails. A canopy of vines shades the building from the intense Texas sun.

History and a Sense of Place

Ecological design is not new. Aboriginal peoples practiced it to enable them to persist for millennia. By necessity, they designed with the climate and with a **sense of place** (J.B. Jackson 1994). Examples are provided by the indigenous domestic architecture that developed in each climatic region. It has only been in the past century that we have ignored the natural limits of place. With the appearance of inexpensive energy, large sheets of glass, and air conditioning, architecture lost its connections with the ancient truth that the Earth dictates the most important building guidance. It was Alvin Toffler in his book *Future Shock* (1970) who commented that the human species had been on Earth for roughly 600 generations and in only the last 2 had there been air conditioning. His point was that, yes, there was life before air conditioning. In fact, human beings first evolved in the hot, dry savannas of Africa and are, therefore able to acclimatize to heat. We keep cool by sweating; our skin temperature is lowered as perspiration evaporates.

A well-known example of buildings designed to heed their environment are the Anasazi cliff dwellings of the southwest United States. These high-mass, adobe-type dwellings were built in south-facing caves, which provided **passive solar** gain in the winter and blocked heat gain in the summer. The local materials for these dwelling were

H-2236 PUEBLO OF ZUNI, NEW MEXICO

Figure 4.1. Pueblo of Zuni, New Mexico. Vintage postcard by Detroit Publishing Co. for Fred Harvey; author's collection.

used in a way that artfully maximized human comfort. In the pueblos of Zuni, Taos, Acoma, and elsewhere in the deserts of the Southwest, their successors use adobe and large mass to create pueblos that make them much easier to keep cool (Fig. 4.1). This thermal mass is critical in buffering outdoor temperature swings. For example, adobe homes in Albuquerque, even uninsulated ones, will stay 15°–20° cooler than the midday peak temperature, often eliminating the need for an air conditioner. In other regions, homes constructed of brick accomplish the same effect.

To duplicate the effect of overhanging cave roofs, houses used to be built with eaves and windows placed so that winter sun would come into the house, whereas summer sun could not enter (Fig. 4.2). This was one of the merits of the California bungalow (Fig. 4.3). These middle-class dwellings were mass-produced from shortly after the turn of the twentieth century until the mid-1920s. With their emphasis on simplicity and practicality, their picturesque use of fieldstone and shingle, and their unadorned structural elements, the design appealed to many home builders throughout the country. It was famous furniture manufacturer Gustav Stickley (1858–1942) who first called attention

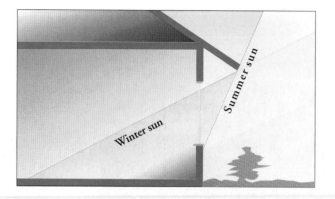

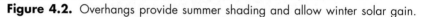

Figure 4.2. Overhangs provide summer shading and allow winter solar gain.

to the bungalow and popularized it in other regions through his magazine *The Craftsman* (1900–1918). This may be one of the reasons they are also referred to as Craftsman-style houses. Interest in bungalows is at an all-time high, both for vintage and newly constructed ones. *American Bungalow* magazine publishes information about both.

Figure 4.3. Craftsman-style house of the 1920s. Vintage postcard by Edw. H. Mitchell, San Francisco; author's collection.

In addition to roof overhangs, bungalows have front porches; the shade from their overhangs cool interior rooms during hot spells and shelter outdoor gatherings. In southern regions, where the ground is swampy and damp, builders developed the tradition of building second-story porches, which wrap around the house and serve as upstairs hallways. Until immediately after World War II, front porches were vital, both architecturally and sociologically, providing an interface between the inside and outside of the house. They were the places where gossip and information about the world was exchanged with neighbors. Today, information comes for other sources, like television. When cars became more popular, many new developments were built to be driven through, not walked in, and architectural styles emerged without front porches.

Front porches are now making a comeback in some new planned communities. However, as noted by Alexander et al. (1977) porches are often made very small to save money, and when they are less than 6 ft deep, they are hardly ever used. In Sienna, a planned unit development in Fort Collins, Colorado, the front porch was among the amenities that the architect provided. In an effort to create the intimacy and flavor of traditional small towns, many planners are trying to create a sense of neighborhood and community. In trying to accomplish this, one of the problems architects have to face is where to place the three-car garages that "grace" the front facades of contemporary suburban houses. They found the solution by placing the garage doors at the rear of the house, facing an alley. Architecturally this may work, but by failing to recognize our well-entrenched consumer habits—many prefer to shop once a week at sprawling discount outlets rather than every day at more expensive corner grocery stores—we will continue to be isolated from our neighbors. We will go to the garage, push the door opener, back out, close the door automatically, and drive to do our errands. Later, returning to our garages and closing the door, we will complete the cycle, all completely isolated from our neighbors.

A different sort of building plan documents these current societal values, observes Akiko Busch in her book *Geography of Home* (1999). The gated communities springing up across the United States look more to the present, not to past tradition. Such high-security suburbs, as they are sometimes called, reflect a different view of neighbors and community than by seen from front porches. Devised and operated by real estate corporations, these communities have a myriad of covenants and restrictions to govern everything from the color of the house to putting up a clothesline in the back yard. (Incidentally, clotheslines are making a comeback in some neighborhoods because they save energy and conserve resources.) It seems that these restrictions, though more about protecting real estate values than preserving American traditions of freedom and good design, are appealing to a growing number of people who find security and comfort in them. Writes Busch (1999, p. 111), "Walls, rather than porches, are the predominant architectural element in many of these communities." Not surprisingly, the kind of casual social interactions that used to bind together communities with porches are avoided entirely. As our homes begin to serve as fortifications, we no longer use the front porch for idle talk and business. She adds (1999, p. 112) that "Alarm systems, walls, and fences have become the 'interface between the indoors and outdoors.' "

There are some general guidelines on building shape and orientation provided in the Rocky Mountain Institute's *Primer on Sustainable Building* (Barnett 1995). Orienting homes to increase the opportunity to use the sun for passive solar heating or daylight can be important in **green design**. In cold climates, building form should be compact to reduce heat loss due to winter winds and slightly elongated on the east–west axis to maximize solar gain. In temperate climates, where the goal is to maximize winter heat gain while minimizing summer overheating, the building should approximate an elongated rectangle running east to west. This minimizes the length of the east and west walls that receive the maximum amount of radiation in the morning and afternoon (Olgyay 1963). The length of the roof overhangs for summer shading is a critical factor; the correct length will vary with climate and latitude. In a hot and humid climate, heat gain through windows should be minimized and ventilation and shading maximized. For hot and dry climates, solar gain should be minimized through shading, especially on the western side. Air movement should be maximized with cross-ventilation.

During the twentieth century, there have been strong movements for ecological planning: William Morris's Arts and Crafts movement, Lewis Mumford's regional planning (Luccarelli 1995), and Frank Lloyd Wright's organic architecture—each celebrated design firmly situated in a wider ecological context. Twenty-five years ago, McHarg produced his seminal text on ecological design (McHarg 1969). Van der Ryn has been teaching ecological design at Berkeley for 30 years. To gain a solid and full understanding of this subject, his *Ecological Design*, with Stuart Cowan (1996), is indispensable. By the way, although successful, the designs tend to be limited, geographically. For example, Wright's Falling Water house (Fig. 4.4) in the Appalachian Plateau of Pennsylvania would not work in the Rocky Mountains because of the heavy snow load.

Successful designs are derived in large part from understanding the natural processes that occur in different regions, and then designing structures and land use accordingly. This approach is the opposite of "one size fits all," in which standard templates are replicated all over the planet with little regard to place; skyscrapers look the same from New York to Cairo. In his book, *The Geography of Nowhere*, James

Figure 4.4. Falling Water. Architect: Frank Lloyd Wright. Photograph by Robert G. Bailey; used with permission of Western Pennsylvania Conservancy.

Kunstler (1993) traced America's evolution from a nation of Main Streets and coherent communities to a land where every place is like no other place in particular, where cities are dead zones and the countryside is a wasteland of sprawling suburbs and parking lots. He makes the case that these are manifestations of our extreme emphasis on private-property rights and reliance of the automobile. Bill Bryson (1990) makes the same case in *The Lost Continent.* As a **roadologist** and veteran of long road trips across America myself, I can sympathize with the boredom he feels. If it were not for the changing geography, it would be hard to tell where you are sometimes; everywhere you see the same tourist junk, fast food, and strip malls. Bryson is rightly outraged at the disappearance of local character and the cheesiness of mass culture. In ecological design, by contrast, the design is sensitive to ecological context and responds to the ecoregion. The task of ecological design is to create land use and structures deeply adapted to place.

In 1939, the Ford Motor Co. exhibit, "Road to Tomorrow," at the New York World's Fair took Americans through a dream world in which teardrop-shaped cars soared along ribbons of concrete high above the farms and forests. In the 1950s, this world was realized in the more elegant stretches of the new interstate highway system then being constructed. Modern design, it was felt, would, before long, solve all modern problems. Ignored at the time were the words of General Omar Bradley: "If we are not careful we shall leave our children a legacy of billion-dollar highways leading to places just like those they left behind."[7] Did he actually foresee our look-alike strip malls and fast-food joints?

All is not lost, however. William Least Heat Moon (1982) in *Blue Highways* reveals a journey taken far away from the "interstates" of the human experience. In this book, Moon gives an account of his 21,000-km journey along the back roads of the United States (marked with the color blue on old highway maps). What he finds is that life manages to persist in ways that it does not in the change-racked "fast lane" into so many of us are swept. In particular, he delights in the small-town cafés and their staple, chicken-fried steak. His way of rating the café he finds is the number of calendars displayed on the walls. The more calendars, the better the café. He warns, however, that if you find one that has more than seven, you better not advertise the fact, as the owners might franchise and ruin it. A classic guide to America's best diners, small-town cafés, BBQ joints, and other very special eateries serving great, inexpensive regional food is provided by Jane and Michael Stern (1992) in their book *Roadfood*.

The downtown Main Street of many small towns in America has been dying a long, slow death ever since the automobile enabled people to shop elsewhere (Hart 1998). They resemble a tree that is still alive but the extremities have "died back." Signs of abandonment are everywhere. Many of the storefronts are boarded up and what businesses that remain usually include a beer joint, video store, and donut shop. Even the ubiquitous Woolworth's store and Midwest café (Fig. 4.5) have closed their doors in many towns. Also, as J.B. Jackson (1984) observes, who has not noticed that in almost every small American town the upper stories of the buildings flanking Main Street are being deserted. Many of the second-story windows in the older brick buildings have been obliterated by commercial facades that were put up in the 1950s and 1960s in an attempt to make the building look modern or up-to-date. Not many years ago, they accommodated the offices of lawyers, dentists, and doctors, dance studios, and insurance agencies. They have moved because the law firm needed more space, the doctor moved closer to the hospital, and the dance studio required more modern wiring. Now all that remains is the gold lettering on the windows (Fig. 4.6). Sooner or later, some of these building will be torn down, to be replaced by one-story buildings or parking lots.

Away from the center of the typical American small town, there is a residential section of tree-lined streets with large houses surrounded by lawns. Where I live, in Fort Collins, they were lived in by prosperous families, the most desirable part of town. Now, many of these houses have been transformed into apartments, especially adjacent to the university where students live off campus. The lawns suffer from neglect and are occasionally adorned by castoff sofas and empty beer containers. The original owners have either died or moved to a more

Figure 4.5. Business district, Laramie, Wyoming, circa 1935. Vintage postcard by Sanborn Souvenir Co., Denver, CO; author's collection.

spacious home on the outskirts of town. Many of these single-family homes are so large that they remind one of institutional architecture. So-called "trophy" homes on large lots, or ranchettes, are totally dependent on the automobile and, thus, contribute to residential sprawl that is unsustainable. Like most American suburbs, they are filled with large, expensive homes, but a larger home is not necessarily a better home. Sarah Susanka (1998, p. 5) says in *The Not So Big House*, "It's

Figure 4.6. Second-story window of the former F.W. Woolworth Co. building in Fort Collins, Colorado showing disuse of the upper story. Author's photograph.

time for a different kind of house. A house that is more than square footage. . . . " The smaller house exchanges space for soul, so that the quality of the space is more important than the sheer square footage, which is meant to impress and seldom gets used. Older neighborhoods are filled with examples of small houses from the past, like the Craftsman bungalow. To meet the demand for such houses where they are scarce, some builders are starting to create neighborhoods full of them. One example, I recently came across is known as Mary's Farm, a planned unit development in Berthoud, Colorado.

A feature article in the October 17, 2001 edition of the *Christian Science Monitor* looks at the appeal of smaller houses. As part of a growing backlash to the "McMansionization" of American housing, according to the article, cottages of less than 93 m^2 (1000 ft^2) offer an appealing alternative to today's bigger homes. In the Seattle area, cottage developments are helping the city stay within urban grown boundaries, yet supply needed housing. They create a sense of community for residents through shared garden space and courtyard design. Across the United States, others are seeing the same sort of potential in smaller-scale living spaces, especially for the nation's population of singles. Builders have found that detached dwellings of 1000 ft^2 or less (less than some suburbanites use to garage their cars) can be aesthetically appealing and highly marketable. Rehabilitation of the many small houses in older, modest neighborhoods could supplement the building of new cottage projects and lessen the damage to the environment.

Retrofitting Main Street buildings has caused a renaissance in Old Town Fort Collins, which, incidentally, served as the model for Main Street in Disneyland, California. Part of the reason is that the embodied energy in an existing building is great, and its reuse will save much of the energy and expense of new construction, eliminating demolition and disposal costs. In many cases, these older buildings have wonderful architectural character that could not be economically replicated. The facades of commercial buildings are being restored and new businesses are clamoring to move in. Redesigning buildings that can have another use at the end of their original life need not mean featureless boxes with undifferentiated interiors at the edge of town. Business is booming because many people like dealing with local merchants. Finding a new use for these old structures has given a boost to the area's economy. Stewart Brand (author of *How Buildings Learn*) says that the issue now becomes how to design so that buildings can evolve gracefully.

There also has been a reinhabitation, dubbed "return migration" by social scientists, as migrants from the suburbs move back to the older neighborhoods. For many people, like myself, the comforts of home are inextricably linked to history. The house I live in was built in 1922. I share Akiko Busch's (1999) feeling about her house, in that I cannot imagine living in a place where no one had ever lived before. Living in such a house also fits my sense of recycling and sustainable use of resources. (I am still driving the Volkswagen Bug I bought *new* in 1964 for $1805.00 or about $1.00 a pound. The odometer shows over 300,000 miles, which is the equivalent of 12 times around the Earth at the equator.) Although this reason is partly true, nostalgic feelings also enter into my values. Although I may rely on high-tech appliances in almost every room of the house, the objects of my greatest affections are traditional artifacts—an old enamel-top kitchen table from the Great Depression era, my great-grandmother's rocker, green glassware (Jadeite), an Arts and Crafts side table, a Navajo rug—that bring a sense of history with them.

My house is located so you do not have to use a car. A neighborhood grocery is two blocks away. The university, library, restaurants, coffee shops, and downtown shopping are within walking or biking distance. Close by are an elementary school and park. These features are attracting others as part of a green movement to renovate and build in Old Town Fort Collins. I recently read an editorial by Robyn Griggs Lawrence (2000, p. 4) of *Natural Home* magazine, in which she writes "that building an eco-house in the country is more damaging to the environment than buying an existing house in town." She goes on to say that ". . . . urban dwellers destroy less habitat, use fewer resources, and generally live in smaller—and more sustainable—homes." She quotes Thomas Schmitz-Gunther (1999) as saying, "To reject the idea of a 'little place in the country' in favor of an apartment in an older, more centrally located building because you don't want to be dependent on a car is to make a very far-sighted and responsible choice." This message is in response to the staggering amount of driving that people do and the horrendous effects of cars on the Earth.

Most people missed this message, as homes in the country and suburbs where one has to jump in the car to buy a loaf of bread are still the most popular. Some, fortunately, are apparently taking this message to heart because the older homes in Fort Collins and elsewhere in the nation are in high demand and are bringing unheard of high prices. The tin or asbestos siding is being stripped off to reveal the original wood facades. Kitchens and bathrooms are being restored to undo the remodeling—referred to as "remuddling" by *Old-House Journal*—of the 1970s that is so common in older homes. Their new owners are scrounging through antique shops and architectural salvage stores to satisfy their passion for original plumbing and lighting fixtures that were hauled to the dump back in the 1950s. Many have hired a color consultant to design an exterior paint scheme that is authentic to the period in which the house was built.

Each region of the Earth has a traditional building form or "vernacular architecture" (J.B. Jackson 1984). Because these styles embody a

great deal of experience, wisdom, and cleverness, it is worth studying
the layout, design, and orientation of older buildings for valuable clues
and ideas. In particular, vernacular architecture is almost always climatically appropriate. Villages in the high-mountain areas where cold
winters prevail are closely packed, the houses drawn together for
greater protection and warmth. Deep overhangs to cope with snow and
rain and storm windows to keep out the cold are typical features of
Swiss Alpine architecture. At the other end of the scale, the human response to heat has resulted in equally regional solutions to the problems of living in climatically difficult environments. The courtyard
house and garden with its cooling plants and water have long been associated with the hot, dry conditions of southern Spain. The specially
constructed wind towers or wind scoops on chimneys mounted on
rooftops in many parts of the Middle East and Indian subcontinent
channel the prevailing wind into every room and maintain more eq-
uitable temperatures than those outdoors. Chimneys themselves have
been designed so that when they heat up in the midday sun, they create an additional through-draft even during calm conditions. In desert
areas, such as Rajasthan, India, this type of ventilation is combined
with moist reed mats that are hung in doorways. Any draft is thus
cooled by evaporation and the result is a natural form of air conditioning. Houses constructed on stilts in Queensland, Australia are designed to promote air circulation. Houses in towns and villages in this
tropical/subtropical desert are grouped close together to shade each
other. Buildings on the campus of Arizona State University in Tempe
are located similarly to achieve the same result. In the Mojave and
Sonoran deserts of the Southwest, arcades along Main Street shelter
shoppers and diners from the blistering sun (Fig. 4.7).

The color of the buildings is one aspect of vernacular architecture
that is frequently overlooked. Roof color, especially, may substantially
affect a building's energy use. In a hot climate, a white or light-colored
roof in combination with well-placed shade trees can lower the building's cooling load by 30%. If one builds roads and roofs of materials
of light color that reflect rather than absorb energy, then it is possible
to cool a city, which, in turn, conserves energy and reduces air pollution. Visionary architect Malcolm Wells (author of *Infrastructure*) says
that some of the same thing could be accomplished by covering public structures—roads, bridges, airports, and so forth—with soil and vegetation. This vision has sparked Ford Motor Company to transform its
Rouge Assembly Plant in Dearborn, Michigan. A vast living floor will
rise above the new plant. Covered with soil and plants, the roof will
become a natural insulator, absorbing rainwater and creating a natural
habitat for hundreds of native species.

Each region of the Earth also has its "vernacular landscape." A good
example as described by Michael Hough (1990) is provided by Tuc-

Figure. 4.7. A covered sidewalk in Boulder City, Nevada provides shade and cooler temperatures. Vintage postcard; author's collection.

son, Arizona. Tucson's desert climate produces a diverse desert-adapted plant community that includes the giant saguaro cactus, creosote bush, and ocotillo. As in Los Angeles (Chapter 2), the blue sky and dry air brought hoards of people to the desert. They were followed by industry and agricultural development that could only be sustained by tapping groundwater from natural reservoirs beneath the desert. In Arizona, the same perceptions about the environment have prevailed as in the California region: the expectation that a dry hot desert can be transformed into a lush, high-rainfall one, while still enjoying the advantages of the dry air, constant sunshine, warmth outdoors and air conditioning indoors.

Until a few decades ago, such an irrigated landscape existed with emerald green lawn and lush trees and shrubs totally alien to the plant communities of the surrounding desert. The landscape today has been transformed. Its once public spaces and private gardens have been replaced with a landscape of giant saguaro, prickly pear, jumping cholla, ocotillo, and many other desert plants set in the gravel and rocky soils typical of the native desert environment (Fig. 4.8). This change in landscape came about through the urgency to conserve Tucson's only

Figure 4.8. A typical suburban house in Tucson, circa 1930. The earlier, high-rainfall front yards of many houses in this region have been replaced by ones like this that utilize native desert vegetation and a drought-resistant design to accommodate the region's soils, climate, and natural beauty. Vintage postcard; courtesy of Susan Sargent.

source of water—underground aquifers. This necessity brought about the introduction of legislation that requires groundwater withdrawal to be balanced by natural replenishment. In contrast, Phoenix's water is derived from three large lakes that collect melted snow from the White Mountains via the Salt River. A plentiful water supply derived from water reclamation and retention from far away has perpetuated its artificial high-rainfall urban landscape, one without context in its region. The recently constructed Tempe Town Lake on the Salt River near Phoenix clearly expresses the different priorities of the two regions in maintaining a level of sustainability with the natural environment. Between the high summer temperatures and low humidity, water evaporates quickly. In fact, evaporation from this 91-hectare lake totals 1889 mm per year, exceeding the natural replenishment rate of 212 mm per year, or nearly nine times as much.[8]

The solution to developing an ecological design grows from its place by integrating design within the *limits* of that place. An example of this approach is provided by a land-capability classification for land-use regulation for the Lake Tahoe Regional Planning Agency (Bailey

Figure 4.9. Land areas in several capability classes may be found around the south shore of Lake Tahoe. R.G. Barton Aerial Photography; annotation by author.

1974). The Tahoe region straddles the California–Nevada border and was established by a bistate compact. Located in the Sierra Nevada Mountains, the basin varies in elevation from about 1830 to 3050 m. It covers about 1300 km^2, of which 38% is covered by Lake Tahoe. Soils in the regions have great erosion potential. Local streams and the lake are extremely susceptible to damage from sedimentation and nutrients. Natural hazards endanger safety and property.

With my Forest Service colleagues Andy Schmidt, Bob Rice, Harry Siebert, and Bob Twiss (also professor of landscape architecture at University of California, Berkeley), I conducted a **land-capability** analysis of the Tahoe Basin in order to develop land development controls that would take into account environmental limitations (e.g., soil erodibility), ecological impacts (e.g., lake sedimentation), and natural hazards (e.g., landslides). The result was a land-use ordinance that divided the region into seven capability classes (Fig. 4.9). For each class, impervious-surface allowances were designed to limit development intensity in sensitive areas (Table 4.1). An inventory of natural system characteristics formed the basis for the land-capability analysis. The characteristics included soils, slope, vegetation, wildlife, fragile resources, natural hazards, and wetlands.

Table 4.1. Land coverage allowances,* Lake Tahoe Regional Planning Agency

Capability district	Land coverage allowed (%)
1	1
2	1
3	5
4	20
5	25
6	30
7	35

*The Land Capability Map identifies the capacities of the lands in the region to withstand disturbance without risk of substantial harmful consequences occurring. These disturbances are expressed in this ordinance in terms of land coverage. Specific permitted amounts of land coverage are established for each capability district. *Source:* Ordinance #13, Lake Tahoe Regional Planning Agency, p. 9
Source: From Schneider et al. (1978).

The Lake Tahoe Regional Planning Agency ordinance was one of the first and best known examples of basing land-use regulations on assessments of the environmental carrying capacity of the land. The use of land cover as a single measure of intensity is simple compared to the complexity of subsequent systems used in other areas. Over the years, the system and map on which it is based have been challenged numerous times by developers—unsuccessfully.

In a related problem, I was to distinguish land capability of a small, local area and its capability within the context of a larger area or region. My solution was to evaluate capability in two ways: on inherent features and limitations of the local area and on the geomorphic features that surround this area. This type of rating excluded small pockets of high-capability lands from development, such as rolling uplands, when embedded in a matrix of highly fragile, erosive, or unstable lands (Fig. 4.10).

New Urbanism

According to a report in the Fall 2001 issue of *American Bungalow*, new small towns and communities patterned after old-fashioned neighborhoods—many featuring bungalow homes—are attracting residents in increasing numbers who are looking for alternatives to residential sprawl. U.S. and Canadian communities developed in the past decade along the principles of New Urbanism, also know as Traditional Neighborhood Design, experienced strong sales in 2000. Although the sales numbers are small, they suggest that the demand for quieter more accessible neighborhoods built to pedestrian scale is still

Figure 4.10. Idealized diagram of a landscape in the Lake Tahoe region, showing rolling uplands (green) incised by closely spaced drainages.

growing. One of the principles of this movement is a distaste for sprawl (i.e., low-density development often far beyond an urban area's edge of service and employment, that separates where people live from where they shop, work, recreate and educate, thus enforcing dependency on automobiles). As sprawl has continued to eat up vast stretches of agricultural land and watershed, the New Urbanists have urged a move toward more compact, mixed-use, economically diverse, and ecologically sound communities, patterned after traditional towns and villages.

Two examples of this objective are Middleton Hills, a neighborhood development in Middleton, Wisconsin, a few miles outside Madison, and Prospect New Town, built on a former tree farm in Longmont, Colorado. Both bear a deliberate resemblance to Craftsman and bungalow neighborhoods of the first third of the last century. They are modeled after the compact American towns built from the 1890s to 1930, with narrow streets, small lots, closely spaced houses located close to the street, mixed residential and commercial uses, and plentiful public spaces. All garages are at the rear of their lots, on alleys that wind through the neighborhood. The Sierra Club singled out the Middleton Hills development last year, holding a press conference at the site to publicize its support for this kind of environmental-friendly development.

Prospect, like Middleton Hills, features narrow, tree-lined streets that connect detached houses, townhouses, courtyard houses, apartments, and live/work lofts to a town center of shops, restaurants, and offices. Builders have erected homes in the bungalow style, as well as more contemporary styles. The developer feels that there is not a strong local vernacular architecture in this part of Colorado. As a result, Prospect is a mix of old and new styles.

Understanding Pattern

Ecological design grows out of an understanding of ecological relationships in a particular region. As we showed earlier, for example, trees that respond to additional moisture that occur along streams are seen repeatedly throughout the semiarid and arid regions of the West (Fig. 3.20). The vegetation of these wet sites is atypical because it exists only because of the presence of the high **groundwater table**—not because of the climate. Other notable examples can be seen where water concentrates around reservoirs, ditches, wetlands, at the toes of slopes, and at the edges of impermeable surfaces, such as rocks, road edges, and sidewalks. The patterns are not isolated objects, but are inextricably linked to the ecological, cultural, and economic processes that created them.

Aspect, or exposure, also affects moisture and temperature, creating topoclimates that result in a regional pattern in the vegetation. Branson and Shown's study (1989) of north- and south-facing slopes in the Denver area describes the contrast. North-facing slopes are steeper than those facing south, owing to differing slope erosion rates. Warmer soil temperatures, increased evaporation, sparser vegetation, and increased runoff contribute to faster eroding and therefore shallower, slopes on the south face. Between the two slopes, drainage courses are often asymmetric in shape and are located closer to the toe of the north-facing side, owing to the south face's debris fill. Trees and shrubs trace the line of north-facing slopes in the foothills and mountains throughout the semiarid and arid regions of the West, especially where slopes are steep enough to differentiate strong topoclimatic contrasts (Fig. 4.11). Shrubs do the same thing along gentler slopes of the High Plains east of the Rocky Mountains.

These topoclimatic contrasts also indicate something about hydrologic processes and landform development. Deep dissection of the mountain range has resulted in steep slopes, large differences in relief, and maximum number of streams. High stream density suggests that water yields occur largely as surface runoff. In this kind of area, sediment delivered downstream as a result of road construction and log-

Figure 4.11. The grass-covered and forested slopes of central Idaho form a repeated ecosystem pattern. Photograph by Kermit N. Larson, U.S. Forest Service.

ging can be expected to be high because of the highly developed transport network (Retzer 1965). Roads must be located in selected, stable areas or designed for stability in unstable areas. The fact that erosion is continuously shaping the slopes and feeding the streams makes it difficult to locate such areas. This is costly but less so than if roads were laid out without a knowledge of the magnitude of the hazards. In steep terrain, logging methods will need to be modified to cause the least possible disturbance to the slopes. For some very steep areas, logging may not be feasible at all. Topography of this extreme kind occurs in the Coast Range Mountains of southwest Oregon and northwest California. In most areas of the Pacific Northwest, road rights-of-way have a substantially higher frequency of landslides than do clear-cut or forested areas (Swanson et al. 1987). In comparison, the gentle topography of the forested high plateaus of Arizona and New Mexico represents problems of extreme simplicity in both planning and management.

Topography and climate also affect biodiversity. Mediterranean regions like Los Angeles tend to have a greater topographical complexity than humid regions because they have more "information" in the form of catastrophic environmental history embedded in them. This complex physical geography, as fire historian Stephen Pyne (1991, p. 62) pointed out, "makes possible, in turn, a complex geography of life, the mosaic of microclimates sustaining a mosaic of micro-biotas." The biodiversity of Mediterranean regions is, in fact, only second to that of tropical rainforests. California alone has more than 7 thousand native plants, nearly a third more than Texas, the next most species-rich state.

The effect of landform on vegetation patterns in the humid regions of the eastern parts of the continents is different from arid regions. Here, instead of striking differences among grasslands and forests caused by landform-controlled topoclimates, the species of forest trees and type of ground vegetation growing on a site will vary as the landform changes (Chapter 3). This creates a diversity of a different kind that is more subtle to the eye and less easy to read.

With increased elevation, the temperature decreases and the precipitation increases generally, creating elevational belts in the vegetation and soils, the character of which depends on latitude (Chapter 3). For example, in the Front Range of Colorado, which lies within the temperate/steppe zone, the belts range from **dry steppe**, to coniferous forest, to mountain vegetation above treeline.

Within the forest cover, the main environmental contrasts in the types of vegetation are not simply related to elevation but to a combination of elevation and topography. We may locate the main forest types on an elevation-topographic gradient. The different types of sites are ordered according to the driest to the wettest conditions. Exposed ridges make the dry end of the gradient, whereas the wetter end consists of deep ravines and flowing streams. Between these two extremes, other sites are arrayed according to their moisture characteristics. By knowing the elevation and exposure, we can predict the kind of vegetation that is likely to occur there. For example, ponderosa pine forests occur on mostly dry, low- to mid-elevation sites, but the distribution varies geographically from region to region (Fig. 4.12). These relationships provide a blueprint for site analysis and planting design using native plants. It spells out the conditions in which plants grow best. Transport the plants to different conditions and problems may occur. For example, quaking aspen from the mountains are most often weakened and stressed in suburban grassland setting, 600 m lower in elevation.

Soils greatly affect the availability of moisture for plants, and distinct vegetation changes often indicate underlying contact points between soil types. With the Great Plains, for example, ponderosa pine [also called rock pine by early botanists; as reported by Woodward (2000)] and shrub islands within grasslands indicate isolate sites of rocky soil conditions, forming reservoirs of water for tap roots

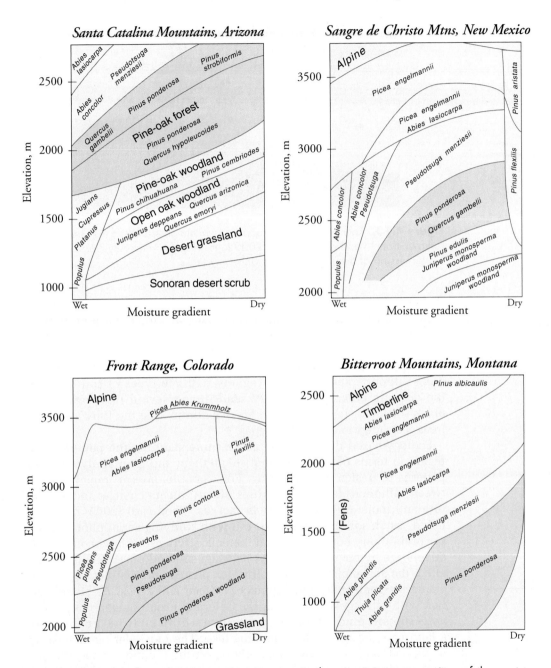

Figure 4.12. Boundaries between ecosystem types in the mountainous ecoregions of the western United States. These boundaries are related to two environmental gradients: elevation and exposure. The green color indicates the range of ponderosa pine forests. Redrawn from Peet (1988); reprinted with the permission of Cambridge University Press.

Figure 4.13. Rocky reservoirs support pines within grasslands of the Great Plains; redrawn from Woodward (2000). ©2000 Johns Hopkins University Press. Reprinted by permission of the Johns Hopkins University Press.

(Fig. 4.13). John Marr (1961, p. 22) wrote of the plant patterns in this region: "Within this area, regional climate is suitable for both grasses and trees; differences in soil and topography, or both, determine which of the two vegetation types will occupy a given locality. Fine, deep soil supports grassland; coarse rocky soil and even fractured rock outcrops support pine." In other regions, extreme types of soil override the climatic effect, such as dry sand dunes devoid of vegetation or black soil over limestone, with grasses dominating the limy soils and trees restricted to other rock types.

Ecology-based design is the act of understanding the patterns of a region in terms of processes that shape them and then applying these patterns to design and planning. For example, abrupt transitions between different plant communities often indicate changes in soil texture and moisture availability. In fact, as Woodward (2000) correctly noted, much soil mapping starts out with studying aerial photos, placing preliminary boundaries, and then classifying soils based on the relatively predictable relationships between soil and plants. Soils greatly affect the soil moisture availability for plants, and distinct vegetation changes signal underlying contact between soil types. As stated earlier, within the grasslands near the Colorado Front Range, rocky reservoirs support pines. Boulders and rocks are difficult to incorporate into a landscape design without looking fake. The most glaring mistake in using boulders is to leave them sitting on the ground without the accompanying plants that they are commonly associated with in the region. Leaving them isolated makes them look as if Bob's Rock Hauling dumped them there yesterday. Placing boulders on the ground is out of place on the Great Plains of Colorado, which are the result of stream and wind erosion and deposition (Fig. 4.14); they would, however, look

Figure 4.14. Evidence of continental glaciation on the eastern plains of Colorado? Note the basement house behind the "glacial erratic." These houses were constructed throughout the West during Depression-era America. The builders had hopes that as economic conditions improved, the above-ground portion would be added. In this example, apparently, they never got around to it. Author's photograph.

natural in Wisconsin, where retreating continental ice sheets dropped large boulders referred to as a **glacial erratic**.

Rocks have a story to tell. Their composition and structure will tell you about the geologic and geomorphic processes that formed them and whether they are suitable for use as ground cover or pavers in a particular region. Lava, for example, is appropriate in regions like the Columbia River Plateau where volcanism is the primary process that formed the landscape. This same kind of rock used in the Great Plains [composed of interbedded sandstone (source of flagstone), limestone, and shale that was laid down beneath a vast inland sea, followed by uplifting and gentle tilting] would look out of place. Similarly, river-worn gravel and cobble is alien in most regional settings. It is mined from the floodplains along major rivers that flow through most major cities. This material—in conjunction with a weed barrier—is used extensively in many yards and parking lots throughout the suburbs of western America regardless of how the natural landscape was formed. On the other hand, developments that are located on alluvial deposits or the extensive alluvial fans at the foot of mountains throughout the arid Southwest and the Great Basin are good regions in which to use this material. It would be a good choice in Los Angeles (Fig. 4.3) or Las Vegas, but not in Denver.

The late landscape architect John Lyle, one of the leading advocates of sustainable design, incorporated the use of local materials in his garden on the outskirts of Los Angeles. He tells the story of his garden in his book, *Regenerative Design for Sustainable Development* (1994, pp. 289–291):

> The garden derives from the larger landscape of which it is a part. The setting is at the base of the San Gabriel Mountains . . . when water from mountain rainfall pours out of the canyons at the mountain's edge, it spreads out and flows at a slower pace over a broad depression called a wash. Washes are important nodes for dissipating floodwaters and recharging groundwater. They are covered with rocks and dotted with plants struggling to take hold. The analogous washes in my garden play a similar role, holding water and allowing it to percolate. The basic material—rock—is the same as that of the natural wash, but it is used in a controlled way at a scale related to home garden dimensions. Thus the form of the garden wash does not mimic the natural wash, but recalls its process in terms related to the human environment.

A remarkable example of the use of local bedrock is found in north central Kansas. Stone posts take the place of wood in this post-rock region, which once had 5 million posts of rough, square-edged limestone (Fig. 4.15). Each post was $5^1/_2$–6 ft (1.6–1.8 m) in length and weighed 350–400 lbs. (159–181 kg). Wood was scarce and expensive in this treeless prairie, which is underlain by a remarkably uniform bed of limestone that is soft enough to be worked with hand tools. One of the advantages of these stone posts was their resistance to the fearsome wildfires that raced across the dry prairie grasslands toward the end of summer.

Another example comes from hydrology. Instead of channeling storm runoff into concrete drains and then to a sewage system, undeveloped drainage swales can be used to mimic nature and help provide sponges for flood protection (Fig. 4.16). By mimicking nature, this system produces beautiful landscaping, reduces off-site water flow, eliminates sewage-treatment concerns, and costs substantially less to build and maintain than conventional storm drainage. In other words, it uses the natural patterns to solve the problem of how to develop the land yet preserve the natural processes.

The best example of this approach to utilizing a natural drainage system is The Woodlands—one of McHarg's most influential projects (McHarg and Sutton 1975). This project is in a 73-km² forest, north of Houston, situated on the flat Gulf–Atlantic Coastal Plain with impermeable soils. It is poorly drained and subject to flooding. A critical concern was how to preserve the woodland environment while draining the land for the new development. Orthodox engineering principles of drainage would have required destruction of the forest in order to lay drain tiles. These principles related to storm drainage were

Figure 4.15. A limestone fence post in the post-rock country of north central Kansas. Sketch by Cathy Johnson. Reprinted from *Kansas: Off the Beaten Path*, 2nd ed., by Patti DeLano and Cathy Johnson. Copyright © by Patti Ann DeLano with permission of Globe Pequot Press, Guilford, CT, 1-800-962-0973, www.globe-pequot.com

Figure 4.16. Drainage swales left natural in an urban development. Redrawn from Barnett and Browning (1995); reprinted with the permission of Rocky Mountain Institute. Original illustration by Jenifer L. Uncapher.

largely developed in the Northeastern cities and were appropriate to the conditions of the crystalline Piedmont on which these cities were located. These principles emphasize accelerating runoff and disposal of this runoff in piped systems.

McHarg reasoned that the Coastal Plain, a great groundwater resource, required the opposite approach, often requiring retardation of runoff to maximize recharge. Ecological design for the Coastal Plain suggested solutions contrary to the orthodoxy of engineering. The elimination of a piped storm-water drainage system as antiecological and excessively expensive has been demonstrated. This project is one of the best examples of ecologically-based new town planning in the United States during the 1970s. Today, the approach might be called "sustainable development." It remains an important model for new community development that is sensitive to the regional ecology.

According to Van der Ryn and Cowan (1996), this idea of observing flows of water over the landscape and designing accordingly is a return to the idea that had its roots in ancient China. Taoist engineers believed that water should meander over the landscape, following its inherent tendencies. By respecting the integrity of existing hydrological cycles, they saw a way of providing ecological flood control. In our own century, we have responded to our own landscapes in a far different way. We have built vast water projects, destroyed wetlands, imposed systems of agriculture alien to the capacities of the land, and mined entire regions beyond recognition. We have not valued landscapes for their own sake, encouraging their own process, instead seeking narrowly productive landscapes that are stripped of their wider ecological significance.

One notable example of the application of Taoist principles can be seen in Wright's Falling Water house. A couple of years ago, I was taking a tour of the house and, near the end, our tour group came to a small alcove. Water was seeping from the face of the bedrock and flowing naturally through the small room before exiting through a small hole in the floor. No attempt was made to stop the seepage, as would be common practice in most buildings in Western culture. Instead, the water was allowed to flow freely and support a small garden of water plants and mosses. What a delightful impression it had on all the members of our group. Our tour leader commented that a few weeks earlier an elderly Chinese man was in the group she was leading. He remarked that when he saw Wright's approach to handling the water, he immediately thought of the Chinese proverb:

> You can't hold back the water;
> You can only direct it.

By working with natural processes, Wright's design is a beautiful application of the wisdom expressed by this proverb.

Ecological methods of flood control are currently experiencing a revival of interest. These methods rely on a healthy landscape in which vegetation moderates flow, erosion is minimized, and water is allowed to follow its own course. The goal is to restore ecosystems so that they can play their long-lost role of controlling flooding. By their design, these natural systems offer additional amenities like recreation areas, trails, and wildlife habitat. Furthermore, they are less prone to catastrophic failures like the great Mississippi floods of 1993.

In addition, say Van der Ryn and Cowan (1996), ecological flood-control systems, although left unvalued in our usual accounting, often have a high monetary equivalent. Wetlands preserve genetic and community diversity and provide food and habitat for migrating birds. Wetlands are nurseries for a wide range of aquatic organisms that would cease to exist without them. They also attenuate floods, purify water, build soil from sediments, regulate groundwater recharge and discharge, and provide local and global climate stabilization. Along the coasts, tidal flats are particularly efficient in purifying water. Only 1000 hectares can filter sewage for a city of 100,000 people. If these services could somehow be assigned monetary value, they would add up to a substantial figure.

The value of the contribution made by natural wetland systems in such processes as water purification cannot be overemphasized. For example, the conventional cleanup of toxic-waste sites is extremely expensive and often causes as much of a health hazard as it prevents. The alternative is to treat the wastes with organisms chosen from the array of living filters. One of the techniques relies on the ability of certain plants to concentrate toxic substances. Cattail is one such plant that typically grows in wetlands. These plants draw these substances up through their root systems, gradually detoxifying the water and soils in the process. The cattails are harvested and then hauled off to a toxic-waste site for disposal. An interesting large-scale application of this process is found in Ford Motor's plan for the new Rouge plant (p. 70). The new plant will use porous parking-lot pavement that will soak up storm water and send it through a biological filtering system before it reaches the Rouge River. Toxin-absorbing vegetation will clean contaminated land.

Leaving vegetation patterns in place can also influence patterns of airflow, moderating pollution. Michael Hough (1994, p. 44) reported that the German city of Stuttgart "has retained the hills that surround the city for parkland and agriculture, because it has found that green hillsides greatly reduce air inversions and pollution problems by maintaining the free flow of katabatic winds that ventilate the city." These winds blow downslope as a result of the force of gravity. Air inversions are a reversal of the normal atmospheric regime (e.g., a temperature increase with height). These inversions inhibit warm air at the surface from rising, thus reducing atmospheric mixing and trapping pollutants. By keeping the hillsides in natural vegetation, they remain relatively cooler with respect to the urban valley and, in the process, tend to restore the normal atmospheric regime of decreasing temperature with height.

Cities have their own microclimates. In general, cities absorb more solar energy and store it better and have fewer plants to cool the air than does undeveloped countryside. As a result, some surfaces in a city can be more than 30°C degrees warmer, and the air a few degrees warmer, than in rural areas. Some 150 years ago, before the Los Angeles region was developed, summer high temperatures averaged about 39°C. By the 1930s, when enormous orchards were planted and fed by water from the Sierra Nevada Mountains, summer highs averaged only about 35.5°C. As freeways and parking lots supplanted groves, however, the summer temperatures have climbed back to desertlike averages. The implication of this is clear, namely, trees can be used to cool cities and houses. Lyle (1994, p. 102) reported that a single tree can "provide the same cooling effect as ten room-size air conditioners working twenty hours per day."

With reference to the general principles involved with using natural patterns and processes in design, it should be noted that Woodward (2000) has raised the need for a caveat. A thorough site analysis should precede pattern application to relate site processes with those shaping regional patterns. For example, the north side of a building provides a cooler **microclimate** in the Northern Hemisphere that is more appropriate for applying plant groupings that are more shade tolerant or are from higher elevations. Climate and landforms of a region should be considered before applying patterns randomly to a site.

Maintaining the Functional Integrity in Ecoregions

When we design ecologically, we observe how regions function and try to maintain functional integrity. Peter Berg of The Planet Drum Foundation (a San Francisco-based group advocating bioregional-level planning), for instance, has observed that the tropical rainforest (Fig. 4.17) provides so much oxygen that it can be considered as a lung of the **biosphere**. He concludes that we should use it not only for massive lumbering, but instead take advantage of its other resources, such as medicines, many not even discovered yet. The Plant Drum Foundation stresses human–land interdependencies, encourages appropri-

Figure 4.17. Rainforest canopy in Papua New Guinea. Photograph by H. Gyde Lund, U.S. Forest Service.

ate technologies, and emphasizes an ethical sense of belonging to a particular region. Manifestations of this movement are the series of bioregional congresses around the country and the magazine *Whole Earth*.

We can create agricultural landscapes that mimic the structure and function of wild ecosystems. At the Land Institute in Salina, Kansas, Wes Jackson and his colleagues have been searching for a sustainable form of prairie agriculture in the image of the wild prairie. In this vision, agroecosystems "should mimic the vegetation structure of natural plant communities . . . and behave like, natural communities."[9] If these structural mimics can successfully recreate their roles in wild ecosystems, then these agricultural ecosystems should inherit many of the ecological functions that lend stability to their wild counterparts.

As Van der Ryn and Cowan (1996) observed, the very pressure that caused prairie vegetation to evolve—an extreme climate ranging from −40°C to 46°C, fire, and large grazing animals—has lent it long-term **resiliency**. The prairie's diverse plant species seek water and nutrients at different soil depths. Their very diversity prevents the spread of diseases and pests. Copious legumes continually add nitrogen to the soil. Nutrients are tightly cycled by the rich soil. The vegetation is adapted to highly erratic precipitation patterns. The perennial ground cover prevents erosion and traps soil moisture. Furthermore, the whole system is well adapted to fire because the seeds of many prairie species require fire for germination. None of these features is maintained in conventional prairie agriculture, with its vast monocultures and dependence on fuel, fertilizer, and pesticides.

An agricultural mimic of the prairie can maintain many of the valuable functions of wild ecosystems while providing respectable crops of food, oils, and animal feed. As a form of agriculture uniquely suited to the land and climate of the region, it works with the underlying ecological processes of the region.

There is a movement to restore our native prairies. Across America, people are setting fire to plots of land and letting formerly productive farm fields go fallow for years on end. These farming practices are the same used by the Plains Indians who had lived on the prairies for thousand of years. They are attempting to reestablish the natural cycles of life on the prairie. People are trying to restore the complex prairie ecosystem by burning away bushes and sapling to allow the grasses and wildflowers of the prairie to return. Farmers are experimenting with growing native crops and leaving some of their land unplowed. Some are willing to put up with animals—prairie dogs, jack rabbits, and coyotes—that have been hunted as pests in the past. Ranchers are letting pastures rest for a season or two to allow plants to grow back, and some are trying to raise bisons, which are native to the prairies.

Changing the natural pattern by adding subdivisions, roads, or other

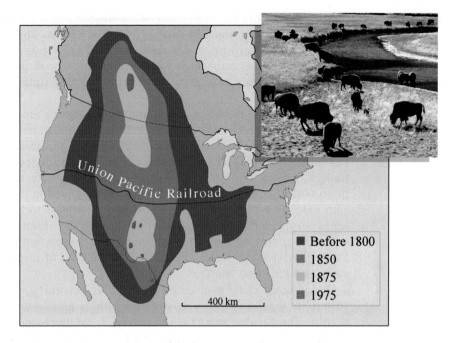

Figure 4.18. Distribution of the bison in North America from 1800 to 1975. Re-drawn from Joachim Illies. Copyright © 1974 by the Macmillan Press; reproduced with permission of Palgrave.

elements changes the functioning of the region. For example, animals change their routes, water flows alter direction, erosion commences, and so on. One of the earliest and best known examples of this is when the Union Pacific Railroad broke the large and intact habitat of the American bison into two patches separated by a corridor (Fig. 4.18). Other scales can be affected too, as when subdivisions and roads are placed between summer and winter ranges of migratory animals.

As discussed in Chapter 2, natural landscape patterns affect the timing of water runoff. Management-created landscapes can have the same effect. An example of this occurs in the Cascade Range in Oregon and Washington. In this region, snow accumulates and melts several times a year, mainly during rain-on-snow events, when water stored in the warm transient snowpacks melts during heavy rainfall. The major floods in basins draining the Cascade Range have resulted from such events. Changing the forest structure in the snow zone by **clear-cutting** increases snow accumulation and also the rate of melt during rain-on-snow events. Warm snow falling in forested areas may catch in the canopy and melt during the snowfall, whereas in nearby clear-cut areas, snow accumulates in a pack. During rainy periods, the snow melts faster in nonforested areas. This suggests that the timing and location of clear-cut areas in drainage basins within the snow zone can affect peak streamflow at drownstream points. A high rate of cutting in the snow zone could have a major impact on peak flows. These impacts might be minimized by distributing cutting units across a range of elevations.

In addition to spatial patterns, temporal patterns are evident within a region [reviewed by Woodward (2000)]. Plants are telltale markers of disturbance and succession. For example, shifts in forest composition can announce the presence or absence of fire and insect outbreaks; although, due to long-term fire suppression, insect-affected forests were more widespread than the fire-affected forests during the past century. Fire patches within the urban/suburban area tend to be small due to rapid suppression. Grazing, another large-scale disturbance, has resulted in the replacement of native bunch grasses with exotic broad-leaved herbaceous plants and dotted landscapes of unpalatable plants such as yucca and cacti.

Overgrazing can have another effect. In Wyoming, grazing has resulted in the native short-grass and bunch grass being invaded by shrubs. The shrubs, being deeper rooted, have increased selenium concentrations in the soils by bringing this toxic element to the surface through leaf fall. Before excessive grazing, the selenium-rich parent material was beyond the reach of the shallow-rooted native grasses. Toxic effects of selenium have been reported in some domestic animals grazing on grasses grown in these now selenium-rich soils.[10] Encroachment is difficult to reverse because selenium quickly accumulates underneath the shrubs, creating islands that discourage grassland recovery.

Maintaining Biodiversity by Leaving Connections and Corridors

Fundamentally, most natural systems are diverse. Therefore, good ecological design will maintain that diversity. Local ecosystems are dependent on the existence of other nearby ecosystems. Therefore, biodiversity depends on leaving some connections and corridors undisturbed.

As Van der Ryn and Cowan (1996) pointed out, sometimes we prefer to imagine that biodiversity can be supported by grossly oversimplified ecosystems, like the clear-cut riddled **old growth** in the Pacific Northwest or a few wild but disconnected national parks and wilderness areas. Sometimes, too, we forget that ecosystems have evolved with characteristic types of disturbance—forest and prairies with wild-fire, floodplains with floods—and that biodiversity hinges on complex patterns of habitat, climate, and renewal. Biodiversity cannot be maintained one species at a time or in one place at a time. It depends on a whole continuum of landscapes, from the fully domesticated to the fully wild.

Local populations of a given species are continually becoming extinct, only to be reconstituted from nearby or more distant surviving populations. The whole landscape plays a role in keeping ecosystems intact.

The basic building block of ecological design to maintain diversity is a *core reserve* off limits to all uses. Each reserve is bounded by a buffer zone with increasingly intensive land uses around it. An elaborate version of the reserve–buffer zone system has been adopted by UNESCO's Man and the Biosphere program, which has developed a global system of almost 300 biosphere reserves. They are located in relation to major ecosystem types (called biogeographic provinces) of the world based on a classification by Udvardy (1975).

At the national level, the U.S. Geological Survey, Biological Resources Division, is undertaking an ambitious biodiversity mapping project know as "**gap analysis**" (Stoms et al. 1998). This project works from layers of geographical vegetation data and known species–vegetation associations to identify unrepresented ecosystem types, habitats for endangered species, or diversity "hot spots." If these areas are unprotected, they are given priority status in the creation of future reserves or wilderness areas.

One example of the application of gap analysis is in the Mojave Desert ecoregion of California (Thomas and Davis 1996). A vegetation map shows 32 natural vegetation types, such as desert grassland. A land-management map was superimposed on the vegetation map that distinguished (1) public ownership with well-developed preservation plans, (2) public ownership with a variety of land uses, and (3) private ownership. The overlays permitted each vegetation type to be characterized as adequately represented, poorly represented, or critically underrepresented in protected areas. Thus, gap analysis identified subregions where ecosystem types are threatened, meaning that there are gaps in regional protection. Rather than focusing on lists of individual endangered species, gap analysis recognizes that preservation of the intact ecosystem is of primary importance. Taking a broad approach may be the most practical way to identify threats and prioritize policy decisions at the regional scale (Franklin 1993).

Wilderness designation continues to be contentious and must be fully justified. An important consideration in setting priorities for additional designations of wilderness is to ensure that underrepresented ecosystems are protected. At the scale of the continental United States, gap analysis has been applied to determine the relative protection currently afforded to different ecoregions. Loomis and Echohawk (1999) found that 23 of 35 ecoregions have less than 1% of their land area protected as wilderness, and 7 of the 35 have no land protected as wilderness whatsoever. Although much of the land with little protection is in areas dominated by private land ownership in the Midwest and Southeast, a surprisingly large amount of land in the intermountain states of Nevada and Utah, which is in public ownership, is substantially underrepresented in the National Wilderness Preservation System as well. Loomis and Echohawk were able to identify ecosystems that should, therefore, be priorities for wilderness preservation recommendations and designations.

However, even carefully selected reserves are not sufficient to maintain biodiversity. Species that have suffered from severe habitat fragmentation may not be able to maintain viable populations without wildlife corridors to connect their small protected areas. Wildlife corridors increase effective habitat areas by allowing safe movement across the landscape. Stream banks, power lines, and hedgerows are commonly occurring wildlife corridors, but they do not meet the need for all species. Deliberately designed and protected corridors provide a crucial kind of connectivity in an otherwise fragmented landscape. A good primer on the subject is *Landscape Ecology Principles in Landscape Architecture and Land-Use Planning* (Dramstad et al. 1996). Wildlife corridors are even more important today in light of global climate change as plants and animals migrate and adapt to changing conditions.

Large mammals are particularly threatened by habitat loss. The grizzly bear offers the most disconcerting example, which is taken from Van der Ryn and Cowan (1996). A single grizzly bear requires about 260 km^2 for its **home range**. Given that 500 or more bears are required for long-term population viability in a single or connected area, this means a minimum of 130,000 km^2 of habitat, far larger than any wild area in the lower 48 states. Clearly, no single wilderness area will satisfy the grizzlies' habitat needs. However, wilderness corridors linking separate patches could potentially create an overall habitat area large enough to support a viable grizzly population.

These principles are being put to use in the proposed Northern Rockies Ecosystem Protection Act (H.R. 2638). The act provides a holistic form of ecosystem protection that explicitly connects several of America's most beautiful wildernesses (Fig. 4.19) and is based on the principle that biodiversity thrives in interrelated ecosystems. Similar wilderness–corridor networks have been proposed for other areas, including the Cascade region straddling Washington State and British Columbia.

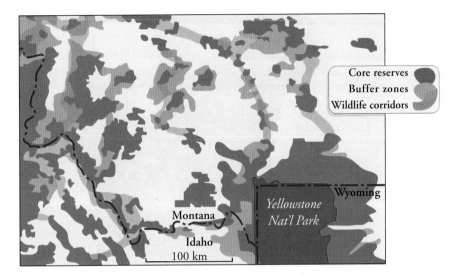

Figure 4.19. System of core reserves, buffer zones, and wildlife corridors proposed by the Northern Rockies Ecosystem Protection Act. Redrawn from Van der Ryn and Cowan (1996).

Honoring Wide-Scale Ecological Processes

Biodiversity implies a diversity of species, but it also implies a diversity of ecosystems and regions themselves. It can be preserved only by addressing all three levels: maintaining viable populations of species, protecting representatives of all native ecosystem types and their successional stages, and honoring wide-scale ecological processes of the regions. These processes include hydrologic cycles, animal movement patterns, and fire regimes. We have discussed animal movement patterns earlier.

In the past, forest fires occurred at different magnitudes and frequencies in different climatic–vegetation regions. As stated earlier, in the boreal forest, for example, infrequent large-magnitude fires carried the flames in the canopy of the vegetation (crown fires), killing most of the forest. Other environments, such as the lower-elevation ponderosa pine forest in the western United States, had a regime of frequent, small-magnitude, surface fires. Here, the burning was restricted to the forest floor and most mature trees survived. Variations of these two fundamental fire types also occurred.

Precolonial fire regimes for different vegetation types in North America have been determined by analyzing fire scars on living trees. In areas lacking trees, the development of vegetation after recent fires, and

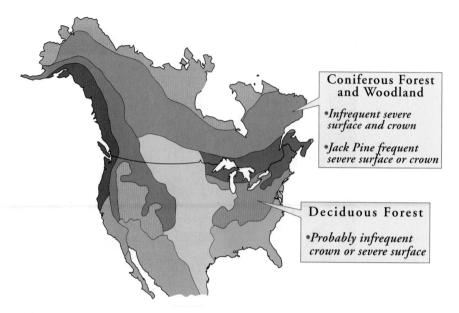

Figure 4.20. Precolonial fire regimes in two different ecoregions in North America. Only major divisions of the ecoregion map are shown. Redrawn from Vale (1982); reproduced by permission of the Association of American Geographers.

early journal accounts and diaries may be used to make an inference about fire regime.

Precolonial fire regimes in the United States are possible to correlate with climatic–vegetation regions (Vale 1982) (Fig. 4.20), such as the ecoregions discussed in this book. A few vegetation types were free from recurrent fires. Tundra, alpine, and warm desert environments had too little fuel for fires. Certain forests of New England, forests in moist topographic situations, and forests in the southern Appalachians apparently were not strongly influenced by fire. Most other forested environments burned with some regularity, although the frequency was highly variable, and both crown and surface fires affected them. Areas with an abundance of herbaceous vegetation seemed to have fire regimes of frequent surface burns.

Where broad-scale landforms break up the ecoregions into different landscape mosaics (Chapter 3) with different **landform relief**, the spread of a disturbance like fire may differ among landscapes. Swanson et al. (1990) hypothesize that in forested, steep-mountain landscapes along the northwest coast of the United States where landform relief does not exceed several tree heights (e.g., Coast Ranges), disturbance agents such as fire and wind can readily move through the forest with little regard for topography. Landforms may have a greater effect on the spread of disturbance and mosaic structure where relief substantially exceeds tree height (e.g., Cascade Range).

In some situations, European settlement increased fire frequencies or intensities. Settlers may have increased fire frequency either through carelessness or clearing forests to encourage the growth of grass.

Changing fire frequency has altered boundaries between plant communities as well. The long-term effect of periodic fire is difficult to judge, but there is some indication that species native to chaparral of the summer-dry Mediterranean climates and the savanna of the winter-dry subtropical climates evolved in association with fire. Preventing forest and range fires may cause changes in the composition and density of the vegetation, sometimes with disastrous consequences when fuels build up. Our suppression of wildfires has extended the intervals between major fire events. These efforts have resulted in fires such as the infamous Yellowstone National Park fire in 1988. This fire not only had a different character from past natural fires, but it also burned a far larger area.

The introduction of non-native species may exacerbate fire hazards. For example, species of eucalyptus trees from southeast Australia were introduced into California at the beginning of the twentieth century. The intended use for these fast-growing trees was for railroad ties and windbreaks for citrus orchards. These tall trees shed their bark profusely; so much so that it accumulates, along with copious leaves, at the base of the tree. The leaves contain oils that are very flammable. When the chaparral-covered hills east of Berkeley and Oakland, California were developed for housing in the 1920s, eucalyptus trees were used for landscaping. It was a disaster waiting to happen—and it did in 1923, and again in 1991 when many homes were burned.

These fires have led to a rethinking of the long-standing federal policy of fire suppression. The Smokey Bear-era of all fire being bad has greatly influenced—and some say hindered—federal forest management. The Park Service and Forest Service have been reluctant to preemptively burn dense underbrush or to allow natural fires to burn out. This policy has created millions of acres of ready fuel—which can be kindled with a single lightning strike. Still, the public has never been comfortable with allowing forest to burn naturally—as happened in Yellowstone National Park in 1988. **Prescribed burns** are also controversial. The 2001 Cerro Grande fire, which destroyed 235 homes in New Mexico, began as a prescribed burn in Bandelier National Monument and was whipped by winds to nearby Los Alamos. Since then, prescribed burns have been suspended or curtailed to give time for a National Fire Plan to be developed and approved. It will call for improved prevention and suppression of wildfires, particularly those near populated areas, by reducing brush and debris that can fuel catastrophic blazes.

The introduction of cattle grazing can affect the regional system by affecting the groundwater table. Recently, for example, the cattle industry moved cattle from the subhumid Flint Hills of Kansas out to the western part of the state, which is semiarid steppe. Corn and soy beans were planted to feed the cattle. Because the rainfall is insufficient to support these crops, deep wells were drilled down to the Ogallala Aquifer to obtain water for irrigation. Prior to this, the streams in the region were **intermittent** or **perennial**. Now, the streams no longer flow, except during and after storms, due to the drop in the groundwater table. The change in the hydrologic processes of the region has changed the regional system, permanently.[11]

The opposite has happened to the South Platte River near Denver. Once an intermittent, mile-wide stream, the character of the river has been transformed by irrigation. Early farmers brought groundwater up to the surface through windmills and wells and then applied this water to new croplands. Water percolated through the sandy, loamy soil, and the South Platte basin water table rose, now sustaining a permanent, year-round flow. Several other modifications changed the form and function of the river as well. **Channelization** of stream banks for flooding and erosion control has shrunk the stream margins drastically. Dams regulated water quantity to reduce flooding potential. Urban development's impervious surfaces increased runoff heading toward the river. As a result, the river is a more stable corridor than before, but the loss of flooding and shifting flow regimes has curtailed riparian forest from colonizing its banks and gravel bars.

Downstream in Nebraska, important changes in channel morphology have resulted from the lowering of discharge caused by flood-control works and diversion for irrigation. Peak discharge and mean annual discharge have declined to 10–30% of their predam values. The South Platte River at Brule, Nebraska was about 790 m wide in 1897 but had narrowed to about 60 m by 1959. The river has changed from a wide braided channel to a narrow and somewhat more sinuous channel.

Wetlands often form on flat or glaciated plains, their sediments producing a rich and fertile soil. Not surprisingly, more than half of the wetlands of North America have been drained and converted to farmland. Throughout the continent, these vital waterfowl habitats, pollution filters, and regulators of flooding have been lost. In several states, including California and states in the upper Midwest, less than 20% of the original wetland area remains.

Some wetlands have been restored from previously degraded sites. In the Prairie Pothole region, which stretches through the Dakotas into eastern Montana, there are thousands of shallow ponds carved into the plains by glaciers more than 10,000 years ago. The potholes provide ideal breeding country for shorebirds, songbirds, and gamebirds that follow the migratory route to Central America and South America. Until recently, farmers freely drained those ponds to expand their fields, planted crops and unwittingly removed much of the nesting cover of North American ducks, 70% of them born and bred in the region. Now, some farmers are paid by the government to turn their fields back into a waterfowl habitat.

Restoring hydrological processes (i.e., the way water flows into and out of the wetland region, and the natural sequence of flooding and drought) may be the most important aspect of wetland restoration. Many wetlands have been altered by diverting water away from them, but once the cycle of flooding and drying is restored, wetland vegetation often follows. The pulsing action of changing water levels aerates the soil and brings new nutrients, greatly increasing productivity. Wildlife soon follows the reestablished vegetation.

The Everglades of southern Florida are a large-scale example of the restoration process. The area became degraded after flood-control measures were taken last century. The government in partnership with conservation organizations are trying to reverse this process by, among other things, returning the dredged and straightened Kissimmee River to its natural condition. A guide for restoring streams and rivers is provided by Rosgen (1996).

In this chapter, we described the patterns and processes of regions and how understanding those patterns and processes can lead to rational land-use patterns that are sustainable. Land development and regional planning certainly will be more successful the more we know about the ecoregional differentiation of a country. Like their component ecosystems, ecoregions are self-regulating and self-sustaining. The concept of sustainable design holds that future technologies must function primarily within ecoregional patterns and scales. They must be based on an understanding of pattern, maintain biodiversity and functional integrity, and honor wide-scale ecological processes.

We next discuss ecoregions as a basis for ecosystem management.

CHAPTER 5

Significance to Ecosystem Management

A 1992 issue of *Bioscience* carried an article by Eugene Odum entitled "Great Ideas in Ecology for the 1990s." In it he states the following: "If we are serious about sustainability, we must raise our focus in management and planning to large landscapes and beyond." In this chapter, I will outline what I have learned to date about ecoregionalism and its significance to **ecosystem management**.

Local Systems Within the Context of Larger Systems

First, by referring to the ecoregion map, the local systems are shown within the context of larger systems. This perspective allows us to understand the interaction between ecosystems at the site level. Processes at the landscape level emerge that were not evident at the site level. Odum (1977) refers to these broader-scale processes as the "emergent properties of systems." The processes of a landscape mosaic are more than those of its separate ecosystems because the mosaic internalizes exchanges among component parts. For example, a snow–forest landscape (Fig. 5.1) includes dark conifers that cause snow to melt faster than either a wholly snow-covered or a wholly forested basin. The conifers are the intermediaries that speed up the process and affect the timing of the water runoff. Understanding landscape processes makes it possible to analyze the effects of managing a site on surrounding sites. We can then assess the cumulative effects that may occur from a proposed activity. Without this understanding, the analyst may conduct a hydrologic analysis of the forest and snow-covered areas sepa-

Figure 5.1. Snow–forest landscape in La Sal Mountains, Utah. Photograph by U.S. Forest Service.

rately and then add the result to erroneously obtain the total runoff from the landscape. The statement "a system is greater than the sum of its parts" applies to the regional landscape as well as individual small ecosystems.

Spatial Transferability of Models

Another application of this perspective is related to the spatial transferability of models. Ecoregions have two important functions for management. First, a map of such regions suggests over what area we can apply the knowledge about ecosystem behavior derived from experiments and experience. We can achieve this without too much adjustment, for example, as in **silvicultural** practices (USDA Forest Service 1977) and seed use (Schubert and Pitcher 1973). Predictive models differ between larger ecosystems; for example, the height-to-age ratio of Norway spruce varies in different climates and therefore ecosystems (Fig. 5.2). In Canada, studies (Huang et al. 2000) have found that the height–diameter models of white spruce were different among different ecoregions. Incorrectly applying a height–diameter model fitted from one ecoregion to different ecoregions resulted in overestimations between 1% and 29%, or underestimations between 2% and 22%. The climatically defined ecoregion determines which ratio to apply to pre-

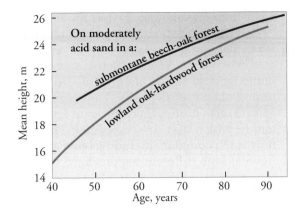

Figure 5.2. Differences in the height-to-age ratio of Norway spruce on similar sites in two different climate-controlled ecoregions. Redrawn from Günther (1955) in Barnes (1984).

dict forest yield. If we select the wrong ratio, yield predictions and forest plans upon which they are based will be in error. We can apply experience about land use, such as terrain sensitivity to acid rain, suitability for agriculture, and effectiveness of best management practices in protecting fisheries, to similar sites within an ecoregion. Second, ecoregions identify broad areas in which similar responses may be expected within similarly defined systems. Therefore, we can formulate management policy and apply it on a regional basis rather than on a site-by-site basis. This increases the use of site-specific information and lowers the cost of environmental inventories and monitoring.

A map of ecoregions would have unquestionable value in identifying types of land that will respond in a uniform way to the application of a variety of management practices. A mapping system for identifying **ecological landtypes** could be useful in current attempts to model the response of wildland areas. Most of these models require that an area undergoing analysis be stratified into homogeneous response units. The ecoregion units described in this book and my previous ones could serve this purpose.

Links Between Terrestrial and Aquatic Systems

As we said earlier, we cannot regard terrestrial and aquatic components of landscape as independent systems, because they cannot exist apart from one another. Just as the lower part of a slope exists only in association with the upper, gullies could not form if no watershed existed. The units of a landscape always comprise connected or associated ecosystems. Within such systems, the diverse ecosystem sites are mutually associated into a whole by the process of runoff and migration of chemical elements. Their common history of development, or genesis, also unifies them. Streams are dependent on the terrestrial system in which they are embedded. Therefore, they have many characteristics in common, including **biota** and hydrology. Rivers and streams flowing from different regions have very different thermal characteristics, gradients, aeration, and biota. For example, the South Platte River in Colorado begins in the high basins of the Rocky Mountains and flows through the rugged Front Range into the Great Plains. This river reacts to different environments with accordingly different characteristics. For instance, the reach of the river in the Rockies supports cold-water fish, whereas the Plains support warm-water fish. Thus, ecoregions make it possible to identify areas within watersheds with similar aquatic environments. A good example of biota that corresponds to a terrestrially defined system is the distribution of the northern hog sucker (Fig. 5.3). This species of fish is widespread but not uniformly distributed throughout the Mississippi River basin. In Missouri, it is

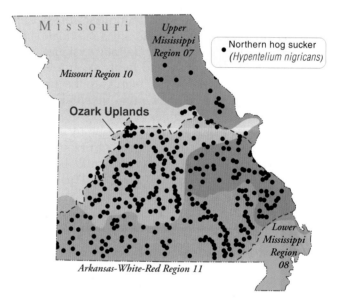

Figure 5.3. Distribution of the northern hog sucker in relation to the Ozark Uplands ecoregion and hydrologic units in Missouri. Fish data from Pflieger (1971); hydrologic unit boundaries from U.S. Geological Survey (1979).

found almost exclusively in the Ozark Uplands ecoregion that covers several watersheds.

A watershed is simply an analytical device based on a single criterion (i.e., topographic control of surface water flow). We need to identify in which ecosystem each part (or all) of the watershed is. That allows us to predict the kinds of stream and associated aquatic organism that will exit there. Figure 5.3 shows the watershed boundaries and one ecoregion. The distribution of the northern hog sucker does not correspond to the watershed boundaries. A comparison, however, identifies areas within the watersheds with similar climatic and landform characteristics and, therefore, similar aquatic environments. These areas are useful for predicting site conditions and for analysis and management of watersheds.

A detailed analysis of the relationship between ecoregions and watersheds is presented by Omernik and Bailey (1997). In the article, the authors conclude the following:

• Considerable confusion and potential mismanagement now exists about the difference between watersheds and ecological regions (ecoregions) for the purpose of ecosystem and resource management.

• Certain ecoregions contain lands that do not easily break down into topographical watersheds (i.e., deep sandy areas, karst landscapes, pothole landscapes, glacial lakes, extremely arid lands, etc.).

• The existing federal designations of hydrologic units often do not coincide with topographical watersheds and often cross ecoregional boundaries.

• Neither ecoregions nor "official" hydrologic units *alone* should be used for ecosystem management.

• Appropriately scaled watershed units *can* be used in conjunction with the appropriately scaled ecoregions to help monitor, define, and comprehensively manage ecosystems and resources. (Omernik and Bailey 1997, p. 946).

What is more, says University of California landscape architect Robert L. Thayer,

> In essence, watersheds and ecoregions in California are 'orthogonal'; rivers and streams cut across zone changes in ecosystems as they flow downhill. This strikes me as a kind of 'plaid', where one must examine both the watershed and the zonal ecosystem to really understand where one is on the 'cloth' that is California.[12]

Thayer goes on to observe that this orthogonal, "plaid" concept really helps to explain the locational zonation of California native peoples, where *language stocks* changed according to major basins (e.g., Patwin in Putah Creek, Nomlaki in Stony and Thomes Creek, Northern Wintun in N.W. Sacramento tributaries, etc.). *Dialects* changed with elevation and, therefore, ecosystem (e.g., "Hill" Patwin versus "River" Patwin, "Mountain" Miwok versus "Plains" Miwok, etc.).

Oxford linguistic anthropologist Daniel Nettle (Glausiusz 1997) has been studying the distribution of West Africa languages by comparing ecological maps showing the ranges of various languages, something no one had done before. He noted, as have others, that languages become more numerous toward the equator. However, Nettle noticed something else: a direct correlation between the length of the rainy season and the number of languages in a region. He reasoned that "if you have abundant rainfall year-round, then you can pretty much produce all the food you need." Contact with the outside world is not essential to survival. However, in areas with more seasonal crops, where failure can bring famine, relations with other groups become crucial. Says Nettle (Glausiusz 1997, p. 30): "So you need to form a social network, which can bring food." The larger the network, the greater the likelihood of a common language.

Terrestrial and aquatic systems influence one another. In steep lands with narrow valley floors and wet climates, the dominant direction of influence is from forest to aquatic system. The composition of aquatic communities and rates of stream ecosystem processes are strongly influenced by the structure and composition of streamside vegetation. The vegetation produces litter and shading, which regulates temperature and light available for instream primary production. The opposite interaction (from stream to forest) may dominate in areas of broad floodplains and especially in arid and semiarid regions where stream water flows outward into the groundwater system, recharging it. Obviously, the point of interchange is also a location where polluted streams may contaminate the relatively clean—and in many cases, pure—water resources in aquifers. These points of interchange are then critical for the management and protection of groundwater resources.

Design of Sampling Networks

Where should we locate monitoring sites? We would like to have detailed information on all of the ecosystems involved in a given study, but that is not feasible because of time and cost. Thus, we must form a sampling strategy that guarantees us, as much as possible, representative information. Two examples of appropriate sampling strategies follow.

Estimation of Ecosystem Productivity

Land management deals with productivity systems (i.e., ecosystems) from which it attempts to efficiently, and continuously, extract a renewable product, such as wood or water.

We need estimates of ecosystem productivity to assess and manage. In order to make such estimates, we must develop the relationships between the ecosystem and the information that we need for production. These relationships may be understood at many levels, from simple judgments based on experience to multivariate-regression and other complex mathematical models. Application of these models is based on the concept of transfer by analogy; that is, we can apply the information gained on monitored production sites to analogous areas. The analogous areas are ecosystems that have been carefully defined and classified.

Such methods are based on the hypothesis that all replications of a particular class of ecosystem will have fairly similar productivity. Some workers have questioned this hypothesis (e.g., Gersmehl et al. 1982) on the basis that correlations between ecosystem types and behavior are generally low.

The reason for this low correlation is that the criteria used to classify the ecosystem types were applied uniformly over an area without considering compensating factors. These factors may produce the same ecosystem type, but for different reasons. For example, soil factors may modify the apparent effects of climate. We know that moisture-demanding plant species often extend into less humid regions on areas of sandy soils because they tend to contain a greater volume of available moisture than do heavier soils. In humid climates, the same soil types support vegetation that is less demanding of moisture than it would be in dry climates. As we will see in the following section, it is unlikely that the behavior of a given type of vegetation would be similar in diverse climates.

One way to establish reliable ecosystem–behavior relationships is to divide the landscape into "relatively homogeneous" geographic regions where similar ecosystems have developed on sites having similar properties. For example, similar sites (i.e., those having the same landform, slope, parent material, and drainage characteristics) may be found in several climatic regions. Within a region, these sites will support the same vegetation communities, but in other regions, vegetation on the sites will differ. Thus, beach ridges in the tundra climatic region support low-growing shrubs and forbs, whereas beaches in the subarctic region usually have a dense growth of black spruce or jack pine. Soils display similar trends, as the kind and development of soil properties vary from region to region on similar sites. These climatically defined regions suggest over what areas we can expect to find the same (**physiognomically**, if not **taxonomically**) kinds of vegetation and soil association on similar sites; see Hills (1960) and Burger (1976) for a discussion of regional differences in ecosystem and site relationships.

As discussed in Chapter 3, the influence of climate on the **ecogeographical** relationships of a region creates unity overall. Such climatic regions delimit patterns of associated aquatic and terrestrial microecosystems that recur over large areas, creating ecosystem regions (Fig. 3.18). Monitoring the behavior of representative sites makes it possible to predict effects at analogous (unmonitored) sites within the same ecoregion. We must carefully select the monitoring sites to truly represent the region. To do so, they must be drawn from all the types of sites found in a region.

Monitoring sites that represent the kinds of ecosystem found in a region will provide more useful information than those selected otherwise. Data obtained from a representative site will be useful for generalizing and applying to unmonitored sites, thereby lowering the cost and time involved in monitoring.

In recent years, numerous publications on ecosystems have appeared. However, only rarely (e.g., Breymeyer 1981, Robertson and Wilson 1985) have they used existing information about the geographic variability of ecosystems to design monitoring programs. Application of this approach requires an understanding of the geographic patterns in ecosystems at varying scales of differentiation—the patterns discussed in previous chapters.

Global Change

Considerable attention has been given to the development of a network of stations for monitoring changes in the global environment (Bailey 1991). Mather and Sdasyuk (1991) suggest two related concepts that should be considered for selecting sites. Obviously, the monitoring site should be representative. Also, stations should be placed where they can detect change. The boundaries between ecoclimatic regions are potentially suitable for this purpose. In such boundaries occur the highest degree of instability of the ecosystems and greatest sensitivity of their components to various forms of pressure.

In cases where establishing new monitoring stations is impractical, existing networks and individual studies have to be used. We can compare existing networks to ecosystem maps to see where representation is inadequate and where additional sites are needed. For example, the Long-Term Ecological Research (LTER) sites are located in various ecosystems throughout North America. By relating the sites to the map, we have a way of establishing priorities for new sites. Furthermore, similar ecosystems occur throughout each map unit. By comparing the

location of the sites to the map, we can see how far the results of research at a particular site can be extended or transferred to analogous
sites within an ecoregion.

Landscaping and Restoration

Each ecoregion has a characteristic pattern of sites. Understanding the
pattern of sites and the processes that shape them provides design inspiration for urban and suburban landscapes that are in harmony with
the region they are embedded within. For example, as Figure 5.4 shows,
desert plants thrive on the arid south side of this Golden, Colorado
house. The north side is moist and humid and is planted accordingly
with mesic species such as spruce. The deciduous **homestead tree**, a
pattern from the 1880s, is placed to provide summer shade but permit
winter light. Where water concentrates at the site's low point, wetland
species are planted that have higher moisture requirements.

Furthermore, like streams, cities do not exist independently of what
surrounds them. Classifying metropolitan areas by ecoregion (Fig. 5.5)
forms a baseline for selecting native plants for landscaping as well as
transferring information among similar cities. In cool and temperate
areas, evergreen trees should be planted to provide windbreaks on the
north and west sides of houses, and deciduous trees on the west and
south sides for shade in summer. In hot, humid areas, deciduous trees
should be planted on the west and south sides of the house for shade
and perhaps on the north side as well. In hot, arid areas, trees and

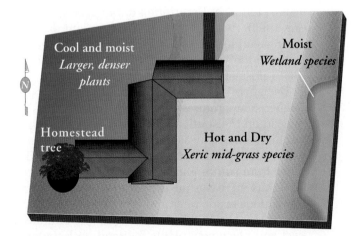

Figure 5.4. Pattern-based landscape design, Great Plains steppe region. After
concepts advanced by Woodward (2000) and Weinstein (1999).

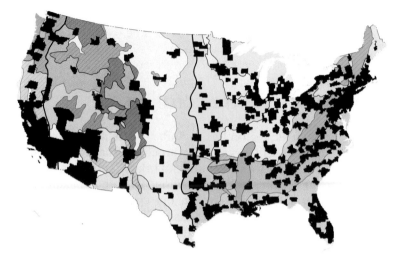

Figure 5.5. Metropolitan areas and ecoregions. Adapted from Sanders and Rowntree (1983).

shrubs should be planted around the house to shade the house walls and create a microclimate and moderate temperatures. Stevie Daniels' (1995) *The Wild Lawn Handbook* has advice for regional wildscaping. Table 5.1 provides a list of tree species that are appropriate for regions around the country. *Landscaping with Nature* by Jeff Cox (1991) lists the most typical native plants—wildflowers, ferns and grasses, and woody plants and vines—for each of nine ecoregions, covering the entire United States. A source of ecoregion native plant information also can be found in *Description of the Ecoregions of the United States* (Bailey 1995). A list of regional native plant experts is included in the Resource Guide in this book. Using this approach, one can quickly characterize the "original and appropriate" ecology of that region and use that information to help guide design.

Designers ignore microclimate at their peril: Mistakes can mean higher temperatures and energy costs. Massive amounts of concrete placed on the south and west sides of a building can increase ambient air temperature by as much as 10°C, significantly increasing the building's cooling load and energy consumption (Fig. 5.6). Also, reflected light from south- and west-facing windows creates high temperatures and thus increases the evaporative loss of water from nearby plantings. Deciduous shade trees and ground covers strategically planted in these exposures can be used to mitigate the situation. Designers should select and place new plants to blend with the existing ecosystem. In addition, they should emphasize the use of native plants and those that are suitable to the region's climate.

Table 5.1. Selected native trees indigenous to ecoregions in the United States

Name of region[a]	Shade	Evergreen
Northern Conifer Forest	Paper birch (*Betula papyrifera*)	Balsam fir (*Abies balsamea*) White pine (*Pinus strobus*)
Eastern Woodlands	Sourwood (*Oxydendrum arboreum*)	Eastern arborvitae, northern white cedar (*Thuja occidentalis*)
Coastal Plain	Sweet bay magnolia (*Magnolia virginiana*) Red oak (*Quercus rubra*)	Southern magnolia (*Magnolia grandiflora*)
Tropical Forest	Mahogany (*Swietenia mahogami*) Twinberry or Simpsons stopper (*Myrcianthes fragrans*) Pigeon plum (*Coccoloba diversifolia*)	
Prairie	Hackberry (*Celtis occidentalis*) Bur oak (*Quercus macrocarpa*) Ironwood (*Ostrya virginiana*)	
Rocky Mountain Forest	Rocky Mountain maple (*Acer glabrum*) Gambel's oak (*Quercus gambelii*)	White fir (*Abies concolor*)
Western Desert/ Great Basin	Big-tooth maple (*Acer saccharum* subsp. *Grandidentatum*) Western chokecherry (*Prunus virginiana* subsp. *Melanocarpa*)	
Sonoran Desert Area	Blue palo verde (*Cercidium floridum*) Arizona sycamore (*Platanus wrightii*)	Desert ironwood (*Olneya tesota*)
Pacific Forest	Mountain dogwood (*Cornus nuttalli*) California black oak (*Quercus celloggii*)	Rocky Mountain juniper (*Juniperus scopulorum*)
Chaparral	Sierra juniper (*Juniperus occidentalis*) Interior live oak (*Quercus wislizenii*) Western yew (*Taxus brevifolia*)	

[a]Regions according to Daniels.
Source: Adapted from Daniels (1995) in March/April 2001 issue of *Natural Home* magazine.

For plants that are suited to the regional climatic context, we need to examine the region's legible plant patterns. Denver's context, for example, is semiarid steppe consisting of native short-grasses and bunch grasses. Urbanization over the last 150 years has created a number of plant patterns that are in response to satisfy the need for comfort and protection from Denver's strong, cold, drying winds and shelter from its summer heat. Many of the plants in this region are native to other

Figure 5.6. The main entry to the Natural Resources Research Center at Colorado State University. Author's photograph.

regions with wet, temperate climates. Many of these patterns are *unsustainable* in Denver's climate. For example, bluegrass lawns are common but depend on imported water, fertilizer, and maintenance. One way to create sustainable landscapes would be to plant a blue grama and buffalo grass lawn. These drought-tolerant grasses are native to the steppe, where they receive 280–360 mm of precipitation annually usually between May and July, sometimes in the form of fierce rainstorms (see Frontispiece). Once established, they do not require irrigation; whereas it is possible for a mesic bluegrass lawn to need over 640 mm of supplemental water per season.

Buffalo grass grows 10–30 cm tall. They spread by stolons (the aboveground stems). They creep across the soil, growing as much as 5 cm a day, and put down roots to grow new tufts of grass in bare spots. Buffalo grasses have seeds that contain germination inhibitors. This is an adaptation to the variability of the precipitation in this region. The grasses will only germinate if climatic conditions are suitable. They will lie dormant for years during a drought waiting for wet conditions to return. As a consequence, plant cover may be difficult to establish without supplemental irrigation at the onset. Once established, the accumulation of too much litter is detrimental to the regeneration of the grasses. The plant cover may become patchy, and **weeds** such as thistle become established. If the steppe vegetation is to be maintained in its original form, a certain amount of grazing is therefore indispensable. This was provided in earlier times by buffalo, antelope, and locust. Nowadays, the grass needs to be mowed occasionally in order to reduce such accumulation.

These grasses, which are resistant to drought, take up water both through their roots and through direct absorption through the blades of grass. They are so good at utilizing the limited water that is available that too much water drowns them. For example, Woodward (2000, p. 57) describes how aggressive irrigated grasses like bluegrass as well as weeds can march right in and take over. "Buffalo grass is not a good competitor for choice spots but very tenacious in marginal locations where juicy grasses dare not tread," says Woodward. This is a problem in urban environments where irrigated water from adjacent property soaks into the soil and flows toward arid sites. This allows the more aggressive weeds and imported grasses to outcompete the dryland grasses. Not only that, there has been a shift in the moisture conditions in urban setting due to irrigation and the growth of trees and structures. This limits the drying effects of winds that define the regional environments of semiarid regions like the Great Plains. As a result, despite our best intentions, the cumulative effects of urbanization limit our ability to replenish native species in urban settings, sometimes resulting in patchy, weedy, unsuccessful attempts. The moral is: Context or surroundings is very important in how a site will operate because the adjacent system affects the site through sheltering, affects on climate, and so forth.

The old saying "if a little is good; then a lot more is better" can lead to failure in attempts to revegetate disturbed sites, especially with grasses. For example, there are many ski slopes in the semiarid mountains of the West. Sometimes, seeds are applied at too heavy a rate in these areas in an attempt to get grasses to come in as thickly as possible to protect the bare land surface. What happens is that the seeds germinate all right but quickly use up all of the available soil moisture. Without irrigation, there are too many plants to survive until the moisture is replenished by rainfall. Consequently, all of the plants soon die and the revegetation process has to be started all over again. The results will be the same unless the climatic conditions of the region are taken into consideration.

In the hot Mojave Desert of southern Nevada, the lawns of Las Vegas drain the city's resources because of the water it takes to maintain them. That is why the Southern Nevada Water Authority has started offering homeowners money to rip up all or part of their lawn and replace it with less water-dependent, indigenous flora. Other states, including New Mexico, California, and Arizona, have instituted similar financial-incentive plans to save water. The National Wildlife Federation offers an official backyard wildlife habitat seal to those who curtail their lawns in favor of native plants that feed and shelter indigenous wildlife and require less pesticide and water.

For years, the American lawn has been cultivated in inhospitable climates from coast to coast and maintained with vast amounts of fertilizer, pesticides, and water. The gardening writer Michael Pollan (1991) in his book *Second Nature* has described such lawns as "nature under totalitarian rule." However, in Las Vegas and other communities, the trend is changing. Whether because of water restrictions, an increased concern about pesticide and fertilizer pollution of streams, or a backlash against the unending labor required to keeps lawns green and mowed, more homeowners are letting their lawns go wild. A source of information about this new approach to lawn ecology is *Redesigning the American Lawn* (Bormann et al. 1993), whose authors criticize the "industrial lawn," composed only of grasses and expunged of any extant weeds. They advocate the "freedom lawn," which allows a diversity of plants to crop up naturally. Such lawns are expressions of regional differences.

Until recently, few designers have explored the underpinnings of regional landscape design. The pioneer in this field is landscape architect Joan Woodward, who has provided guidance on applying regional sensibility to landscape design in the Denver Front Range (Woodward 2000). Another landscape designer working to achieve the same objective is Judith Phillips. In her book, *Natural by Design* (Phillips 1995), she presents a thorough treatment of landscape design, planting, and maintenance in the ecosystems of the arid southwestern United States. It is based on an understanding of why and how plants grow where they do within the region. Landscape meaning and regional thinking are an important part of an ecosystem approach to home gardening, say Mark Francis and Andreas Reimanns (1999) in their book *The California Landscape Garden.* They used California's natural beauty and habitat as a starting point for inspiring Californians to see their gardens as extensions of the surrounding landscape and responsive to the various regions of the state.

Woodward calls her approach **pattern-based design**. This approach is distinguished from another popular design approach called **Xeriscape**.[13] Xeriscape design, or water-conservation design, involves zoning plants for microclimate control, reducing turf areas, and reducing **evapotranspiration** through the use of mulches and efficient irrigation. Xeriscape design promotes the "greening" of semiarid garden landscapes while encouraging water conservation. This is the principal goal; designing for regional distinctiveness is not a stated goal. The pattern-based design differs from Xeriscape by incorporating an understanding of patterns and by remaining sympathetic to regional processes.

There are a number of public Xeriscape demonstration gardens throughout the West that are designed to educate and convince the public that water-conserving gardens can be lush and beautiful. There are also dozens of gardening books that illustrate Xeriscape principles. These gardens and books, however, are purposefully generic and apply to many design styles. Substituting drought-tolerant plants seem to be the favored model in the demonstration gardens and books. As far as I know, Woodward is the first to build on Xeriscape's useful resource-conserving foundation and offer guidance in creating regionally distinctive landscapes: those based on a region's legible plant patterns.

One aspect of landscaping that is too frequently overlooked is food production. As Barnett (1995, p. 37) writes,

Today, most food consumed in this country is grown with chemical fertilizers and pesticides, freighted over long distances and then sold by huge supermarket chains. By some estimates, it now takes ten times as much energy to grow, ship, and process our food as the food itself contains. Apples from New Zealand and grapes from Chile look appealing in the grocery store, but this energy-intensive agriculture is, in the long run, inherently unsustainable.

For this reason, a garden, orchard, cropland, or some other form of edible landscaping should be incorporated into almost every green building project. The home garden is more popular than ever. However, edible landscaping can be used on commercial sites as well. To boost productivity, one should use organic, permaculture methods that blend fruit and nut, grapevine, and other edible plants and shrubs with ground crops (Mollison 1964). Like garden books, the published sources which offer advice on how to accomplish this kind of landscaping tend to be generic and not specific to an ecological region. A friend told me about a person she knew who was very successful with permaculture methods in central Kansas, a temperate steppe-type ecoregion. This person was hired to work in applying the same methods in Hawaii, which is a humid tropical type of ecoregion. After several years of unsuccessful attempts, the project was abandoned. One wonders if the reason for lack of success could be that the method was not modified to fit the changes in ecological conditions between Kansas and Hawaii. Another factor that may account for this result is that the methods were applied to disturbed land in Hawaii, whereas undisturbed land was used in Kansas.

Gardens and edible landscaping are easily integrated into a landscape. For example, grapevines can be used to screen outdoor sitting areas (Fig. 5.7). In addition to providing food, the vines become part of the cooling system. They can be set on a trellis or **pergola** attached to the house or next to it. A built-in or detached green house can extend the growing season, providing fresh, pesticide-free vegetables during early spring and late fall. Also, of course, as Barnett (1995) reminds us, food production is not limited to plants. Indoor or outdoor fish

Figure 5.7. Deciduous grapevines can provide summer shading and allow winter solar gain. Author's drawing.

ponds can also be incorporated into landscape design. In some cases, a fish pond may even serve as the last step of a wastewater-treatment system.

Like all microclimates, gardens are affected by hills, hollows, and orientation to the sun. A fence built on a low spot on the northeast side of the property is going to impede the flow of evening drainage air, which makes it no place to plant tomatoes; in most places, tomatoes like to be planted next to light-colored walls, the better to absorb radiation. In medieval European gardens, you see fruit trees planted right next to walls, because the people knew that walls radiated heat, which led to better fruiting. In regions with cold winters like Colorado, however, fruit trees planted on the sunny side of a house will be less likely to bear fruit because that side of the house is warmer and the trees with begin budding too soon. Later, they are more likely to be affected by late-spring frost that prevents the trees from bearing fruit.

Nature writer Kevin Cook offers advice on how to successfully attract wildlife to the garden. He says that "to develop a really successfully wildlife garden, you must identify the wildlife possibilities for the ecoregion in which you live and then arrange your garden to attract those species unique to your area."[14] Here is how the idea works: Planting berry-producing shrubs is a vital wildlife gardening strategy, because berries will attract both adaptable species found across ecoregions (e.g., American robins) and unique species found in particular ecoregions. Berries attract varied thrushes in the Pacific coast ecoregions, bohemian waxwings in the Rocky Mountains, phainopelas in the deserts of the Southwest, brown thrashers in the Great Plains, and so on. Thus, one strategy yields different results, depending on the ecoregion.

Regional Environmental Problems

Society needs a way to estimate and manage the impacts of environmental problems that affect large geographic areas. For example, hazards such as **landslides** occur extensively in certain regions (Radbruch-Hall et al. 1982), such as northwestern Wyoming (Fig. 5.8), creating a regional problem. In this region, soft bentonitic claystone and shale of Jurassic and Cretaceous ages are particularly susceptible to **slump-earth flows** and **debris flows.** Shale, siltstone, and sandstone of the early Tertiary age also slide, especially where overlain by massive volcanic rocks. Shale overlain by massive rocks is particularly unstable where the rocks dip toward a valley and the massive rocks have been cut free by erosion. The lower Gros Ventre slide east of Teton National Park is

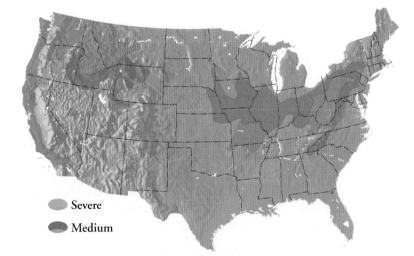

Figure 5.8. Severity of landslide problems in the United States. Adapted from Swanston (1971).

Figure 5.9. Looking west over the 1925 debris flow and resulting lake on the Gros Ventre River, circa 1971. The top of the white scar is 640 m above and 2.4 km back from the river. Author's photograph.

a well-know example, in which 38 million cubic meters of Tensleep Sandstone slid down over the underlying shaly Amsden Formation in 1925 after a heavy storm (Fig. 5.9). The rubble was carried 107 m up the opposite side of the valley and the main mass formed a dam 76 m high. The lake that immediately began to form behind the dam was eventually 61 m deep. Two years later, the lake overtopped the dam and discharged over 5300 hectare-m of water. The resulting wall of water raced down the valley, did much damage, and, despite warnings, cost the lives of several householders who were caught trying to save some of their belongings. Driftwood lodged in trees 3 to 4 ft above the valley floor. As a channel cut through the dam, it also became longer, which retarded the downward cutting. At the same time, the lake was shrinking and seepage continuing, so that a balance was soon reached. Little change has occurred since, although from a geological point of view, the lake is temporary because of the inevitable infilling by sediments and slow lowering of the outlet by erosion. In fact, there is geological evidence that the river had been dammed previously. Lake deposits found 2.4 km upstream from the dam have been dated by carbon-14 and have yielded an age of 4120 ± 200 years.

This is one of only many landslides in this valley. Of possible significance to dam construction in the Teton region was the reaction of old landslide areas to the lake formed behind the debris-flow dam. As the new lake began to fill, many of the toe areas adjacent to the lake became unstable when the cohesion and strength of the landslide deposits were reduced. A reactivation of an old landslide toe occurred 6 days after the lower Gros Ventre slide. Saturation alone was not the cause of all sliding associated with the newly formed lake. Equally important was the rapid drawdown of the water level in the lake after the dam failed. The lowering of the water table left saturated slopes and landslide deposits above the lake surface without the additional stabilizing effect of the weight of the water against the slope. Groundwater moved toward the slopes and seepage pressure developed that made the slopes less stable. As a consequence, numerous slides developed between the original high level of the lake and the present water surface.

The other geographic factors creating conditions that make slopes susceptible to failure in this region include high relief, steep slopes, deformed weak bedrock, earthquakes, abundant precipitation, and stream undercutting. Building and road construction in such regions may remove critical support, destabilizing the slopes. Also, by removing the vegetation for timber harvesting may change the hydrology of the slopes so that they become wetter and more prone to movement. Slides may occur soon after logging or occur years later when the binding roots have decayed. Such landslides contribute increased debris to streams, encouraging both flooding and a serious deterioration in water quality. Decreased water quality adversely affects the biota living in the streams and lakes downstream from the slides.

By knowing the geographic factors that cause slides within a region, one can identify and avoid hazardous landslide areas during road construction and logging operations. Vegetation and topographic anomalies within a region are useful for identification of those areas. For example, in the Middle Rocky Mountains on slopes between 1800 and 3200 m elevation, dense aspen growth, in an area normally supporting evergreen forest, indicates the location of earth flows (Fig. 5.10). Earth flows do not support slow-growing conifers; instead, because of soil movement, fast-growing aspen replace them, indicating wet ground conditions. Aspen may be present because they reproduce mainly by root suckers. Soil movement disturbs the roots, stimulating sprouting and probably contributes to aspen spreading. Continued movement shifts and tilts the trees (Fig. 5.11).

In Teton National Forest in the Rocky Mountains of Wyoming, for example, I used this approach, with the help of geologist J. David Love

Figure 5.10. Vegetation anomalies reveal unstable slopes in the Rocky Mountains.

of the U.S. Geological Survey (Fig. 5.11),[15] to develop a **slope-stability** map (Fig. 5.12) with three major classes: unstable (16.2%), potentially unstable (41.2%), and stable (42.6%). This map was used in a study of **commercial forest land** to determine which lands should be excluded from future timber harvesting. Severe limitations make the unstable land unsuitable for logging, road construction, or other soil disturbance, and, therefore, these areas have been classed as unusable. Potentially unstable slopes are classed as potentially usable; that is, they can be used with improved technology and with more detailed land-use data to identify limits of use. Land disturbance should proceed with caution in these environments. A map of this sort can be used to avoid landslides altogether or to minimize the effects of disturbance by taking appropriate precautions. These maps provide information to assist engineers and land managers in estimating the extent of impending landslide action for broad-scale planning purposes. Furthermore, a knowledge of the location, characteristics, and susceptibility of landslide deposits will aid in utilizing these areas with minimum disturbance and costs. Personnel of the Ottawa National Forest in Michigan, for instance, use maps of ecological land types to locate roads for less cost in the Laurentian Mixed Forest ecoregion.

Figure 5.11. During a field verification of slope-stability mapping of the Teton National Forest in 1970, geologist J. David Love poses in a grove of shifted and tilted aspen trees on Mud Creek mudflow, upper Gros Ventre River valley, Wyoming. Author's photograph.

In Chapter 3, we discussed how the character of landscapes is vastly different in different climatic regions. This is because the magnitude and frequency of geomorphic processes varies by the climate region. Rocks weather and erode differently in cold regions than they do in warm, arid regions. Some regions are older, geologically stable, and climatically predictable. In New England and the Midwest (temperate forest and grasslands), energy flows through the environment in a seasonal pattern that varies little from year to year. Geology is generally quiescent and natural processes are rarely catastrophic, except for killer tornadoes. Frequent rainfall of low and moderate intensity is the principal geomorphic agent, and the landscape is composed of gently rolling slopes well covered with vegetation.

Southern California, on the other hand, is characterized by high-intensity, low-frequency events—flash floods, catastrophic fires, and monumental earthquakes. This region is on the edge of the North America crustal plate with active mountain building and associated vol-

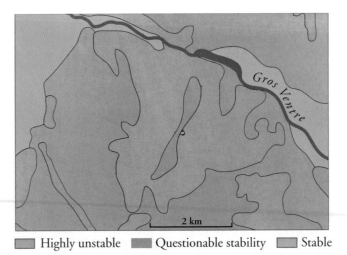

Figure 5.12. Portion of slope-stability map, Teton National Forest, Wyoming. Adapted from Bailey (1971).

canism. High-relief, stream-dissected mountains are interspersed with basins filled with debris eroded from the mountains.

In northwest Wyoming, the dominant geomorphic process that shapes the land and provides hazards for land use is **mass movement**. In other regions, such as Southern California, summer droughts followed by winter storms, create catastrophic floods (Fig. 5.13) and **soil slips**, especially when the vegetation is removed by wildfire. This was not a problem until the land use in the valleys surrounding the mountain ranges began changing. For example, before the 1920s, the San Fernando Valley, 16 km northwest of downtown Los Angeles at the foot of the San Gabriel Mountains, was a land composed of small towns surrounded by citrus orchards, groves of avocados and walnuts, and chicken ranches. This all changed during the huge population explosion of the 1920s when the land started to be converted from agriculture use to suburbs. Residential developments were located at the mouths of the canyons coming out of the mountains and along the many floodways, or washes, flowing across the alluvial deposits composing the valley floor. All was well until 1938, when a week-long storm hit the region and created a great deluge. That year, an unprecedented 87 people lost their lives and 121,406 hectares were converted into an inland sea. My parents, Gale and Barbara, were living in the Valley at the time and were flooded out of their home, but they survived. Personally speaking, I think that was fortunate for me because I was born the following year.

Figure 5.13. Flood damage near La Crescenta, California, circa 1938. San Gabriel Mountains in background. U.S. Forest Service, unknown photographer.

In tranquil New England, rivers swell to flood with a mere doubling of their normal flow. In southern California, the Los Angeles River—usually a sluggish stream—has been known to increase its flow 3000-fold in a single 24-h period. Local erosion and sedimentation rates accelerate explosively. What distinguishes this region from other American environments is the contrasting roles extreme events play in each. In this region, "disasters" are the ordinary agents of landscape and ecological change. Under this climatic regime, storms of 50-year or 100-year occurrence accomplish the most geomorphic work. In regions like this, Mike Davis (1998, p. 18), the author of *Ecology of Fear* points out, "The extreme events that shape the Southern California environment tend to be organized in surprising and powerfully coupled causal chains." Drought, for example, dries fuel for wildfire that, in turn, removes ground cover and makes soils impermeable to rain by making them hydrophobic.[16] This increases the risk of flooding in areas where **tectonic activity** is constantly exposing new surfaces to erosion and increasing stream power by raising land elevation. In such conditions, storms are more likely to produce dramatic erosion and landform change. He goes on to say, "Vast volumes of sediment rapidly realign river channels, and before the advent of twentieth-century flood-control engineering, even switched river courses between alternative deltas. Sedimentation can also create sandbars that temporarily cut off tidal flows to coastal marshes—initiating a 50- to 75-year long cycle of ecological readjustment."

The combination of wet winters and dry summers is unique among climate types and produces a distinctive natural vegetation of hard-leaved evergreen trees and shrubs called **sclerophyll**, scrub woodland, know as *chaparral* in California. This type of vegetation reduces water loss with leaves that are small, thick, and stiff, with hard, leathery, and shiny surfaces. Extreme flammability characterizes this woodland during the long, dry summer. This poses an ever-present threat to suburban housing that has expanded into chaparral-covered hillsides. Prior to city sprawl into the hillsides, the chaparral would burn on average every 30 years or so (Fig. 5.14). When an attempt is made to keep fires from starting, the brush grows thicker. As a result, the brush fires are more destructive, particularly when homes are built in wildfire corridors. Fire, fortunately, does not kill the chaparral whose deep roots anchor the soil to the steep slopes. These plants sprout back from root crowns (Figure 5.15). The seeds of some species need fire to germinate and may lay dormant for years until the next fire.

One needs to recognize, as Davis (1999, p. 100) says,

> . . . that the region's extraordinary fire hazard is shaped, in large part, by the uncanny alignment of the canyons with the annual "fire winds" from the north: the notorious Santa Anas, which blow primarily between Labor Day and Thanksgiving, just before the first rains. Born from high-pressure areas over the Great Basin and Colorado Plateau, the Santa Anas become hot and dry as they descend avalanche-like into Southern California. The San Fernando Valley acts like a giant bellows, sometimes fanning the winds to hurricane velocity and they roar seaward through the narrow canyons. . . .

Once ignited, fires are so intense that firefighters can do little while waiting for the winds to abate or fuels diminish. This is a good example of how one regional ecosystem, hundreds of kilometers away, is linked to another and affects it dramatically.

The dry-summer, winter-wet climatic region of southern California lies hemmed in between high mountains and the sea. The surface features include small, isolated valley lowlands, bordered by hills and backed by high mountain ranges. When fire burns the cover, the torrential winter rains that follow produce mudflows and floods. The soil is swept completely away, leaving only the bare rock exposed at the surface, and bordering valleys are filled with mud. The situation is worsened in southern California, where soil becomes water repellent

Figure 5.14. A pre-1912 fire in the San Gabriel Mountains north of Los Angeles. Vintage postcard by Edward H. Mitchell, San Francisco; author's collection.

following fire. The extreme fires that occur in this region can transform the chemical structure of the soil itself. The volatilization of certain plant chemicals creates a water-repellent layer in the upper soil, and this layer, by preventing percolation, dramatically accelerates subsequent sheet flooding and erosion.

The population continues to grow in many urban and suburban areas like Los Angeles. Most of the lowlands and valleys have been developed for many years so that houses are now being built on the

Figure 5.15. Chaparral sprouting after a fire in southern California. Photograph by U.S. Forest Service.

surrounding hills and the lower mountain slopes. Removal of the chaparral for housing and **fuel breaks** may exacerbate the erosion problems (Fig. 5.16). Thus, we need to understand the role that chaparral plays to avoid ecological mistakes.

Detailed studies of the San Dimas Experimenal Forest near Glendora [reviewed by Bailey and Rice (1969)] have demonstrated that the frequencies and areas of soil slips are three to five times greater for grass-covered slopes than for brush-covered ones. These studies also found that the minimum angles for failures were less for grassy cover than from most chaparral vegetation. The roots of grasses are typically shallow and spread laterally in the shallow **colluvial** soil instead of down into the bedrock. Chaparral-covered slopes, on the other hand, are characterized by species that have tap roots that extend into bedrock fissures for great depth. Oak chaparral, one of the vegetation types least susceptible to slippage, is composed primarily of California scrub oak with roots extending to depths of 9 m. Soils are susceptible to increased slippage if the brush is converted to grass, especially after 7 years, when the anchoring roots have decayed.

These studies have shown that all slips occurred on slopes greater than 80% (about 39°) in this region. This suggests a means for identifying areas with the greatest hazard from slope maps or topographic maps, to determine areas that should be maintained in brush. This is not only of importance to sustainability but also to public safety. Southern California residents have suffered death, injury, and property damage from debris flows generated by soil slips that occur during heavy rains; the process is a recurring major natural geomorphic agent in the

Figure 5.16. Massive soil slip in Griffith Park near downtown Los Angeles, circa 1969. Author's photograph.

region. During the years 1962–1971, 23 people in the greater Los Angeles area died as a direct result of being buried or struck by debris flows that probably originated as soil slips (Campbell 1975).

The regional landscape of southern California is related ecologically to mid-latitude regions of hot, dry summers and mild, wet winters (the classical Mediterranean, central Chile, and the coastal zones of South Africa's Cape Province and West and South Australia). Comprising but 3–5% of the Earth's land surface, these Mediterranean ecosystems are the rarest of major environmental systems. Similar climates acting upon comparable tectonics have produced torrential landforms and erosion patterns, as well as equivalent frequencies of floods, landslides, and earthquakes.

Southern California illustrates the idea that climate influences geomorphic processes, and, hence, landscapes. In the field of **climatic geomorphology**, these climate–process systems have been called **morphogenetic regions** (or morphoclimatic regions) in which distinctive groups of processes operate. Table 5.2 summarizes both the climatic and the geomorphic aspects of the major systems. This table includes

Towns on the Plains would serve as centers to service a new industry offering tourist safaris to a new national playground. For small towns like Arthur, population 115, in the Sand Hills region of western Nebraska, this is not viewed as good news. Certain towns would be given priority for public funds over others because there is not enough money to go around. This dilemma has generated another notion that is receiving serious attention in some quarters. It goes by the ominous name of "triage," the process used by the French in World War I to select for medical treatment only those wounded soldiers who appeared to have a good chance at survival. As public policy, triage could seal the fate of places like Arthur by concentrating funding on other regions that can better serve as centers for growth.

Whether or not the Plains are returned to their preagricultural condition, they are inescapably threatened by the world's changing climate. According to computerized global climate modeling, the Plains are one of several regions around the world that will turn into a major desert if the predicted CO_2-induced **greenhouse effect** takes effect. Farmers will have a harder time responding to this **desertification** because it would take three times more water than today's rate to compensate for global warming. This would seriously harm most strategies for survival in this region.

We used climate as the basis for recognizing global ecoregions. By studying climate-controlled ecoregions, we gain an understanding of the geomorphic processes that shaped and continue to shape the region. These formative processes are reflected in the patterns we see repeated throughout the region. These patterns are what William Marsh (1998) calls "fingerprints of the formative process." These patterns, interpreted in terms of process, are the point of departure to evaluate questions of planning for expanded exploitation and management of natural resources.

Transfer Knowledge

Analogous regional-scale ecosystems occur in predictable locations in different parts of the world because the controlling factors are the same for each. The middle-latitude combination of continental position, cool winters, hot summers with rainfall during the summer, and snow during the winter, for instance, gives the hot continental conditions that may be recognized not only in the northeastern United States but also in parts of eastern Europe, northern China, Korea, and central Japan (Fig. 5.17). This makes it possible to transfer knowledge gained from one part of a continent to another and from one continent for application in another. International Seeds, Inc., for instance, has found ex-

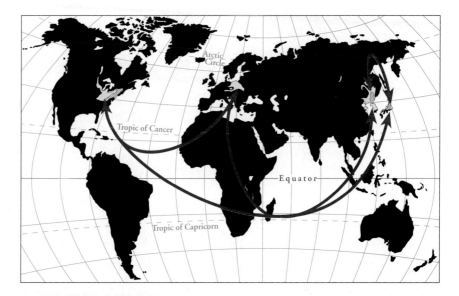

Figure 5.17. Data gathered from one ecoregion can be applied to another of the same type, where there is little or no data, in this case, the hot continental ecoregion.

cellent agronomic predictability among similar ecoregions around the world. This principle is illustrated in Fig. 5.18. I took the photograph on the left side of the figure of the hot continental, broad-leaf forest in Pennsylvania. My son took the photo on the right in the same type of ecoregion in Japan. Each of these regions would have the same processes and problems.

By examining plants associated with similar ecoregions in other parts of the world, we can expand significantly the number of species available for applications in our ecoregion. Again, however, there is a warning: Care must be taken to not introduce species that can escape from cultivation and take over habitats from native species. Woodward (2000) describes how tamarisk, or salt cedar, has taken over many riparian niches along western bottomlands in the southwestern United States. These shrubs send roots to the water table and are so efficient in tapping the water supply that they lower the groundwater table levels significantly. Their deep roots also extract salts from deeper layers of soil and deposit them through salt-excreting glands at the soil surface. Less salt-tolerant riparian species such as willows and cottonwoods cannot compete in such hostile soil conditions and are declining in numbers. Streamsides have also been invaded by Russian olive, which was imported from the steppes of Russia. This species has replaced cottonwoods and willows, which generally add to species di-

Figure 5.18. The hot continental, broad-leaf forest of Pennsylvania, USA (left) and Honshu, Japan (right). Photographs by Robert G. Bailey (left) and Matthew G. Bailey (right).

versity in semiarid regions. They provide a versatile habitat—good insect breeding for birds, good cavity formation for cavity dwellers, good nest sites, shade for fish species, and understory woody and herbaceous plants. Russian olive on the other hand, is so hard that cavity nesters cannot build in it and it harbors few insects. The tree grows in the shadows of the cottonwoods; once the native trees have died, the olives monopolize all habitats, reducing the likelihood of natives establishing themselves. They, thus, threaten riparian habitat diversity throughout the West.[17]

Caution is necessary not only in transferring data on adaptable species from one continent to another but in transferring such data from one region to another, even when the same species are involved. Species with wide geographical ranges (e.g., sagebrush) often develop locally adapted populations, called **ecotypes**, having different limits of tolerance to temperature, light, or other factors. Compensation along a gradient of conditions may involve genetic races or merely acclimatization. The possibility of genetic variations has often been overlooked in applied ecology, with the result that restocking or transplanting has often failed because plants from remote regions were used.

Another ecoregion is the shrub grass steppe that extends from Canada into Colorado and from the Rockies east to central Nebraska. The steppe is characterized by semiarid climate, shrub grasses and brown or chestnut soils. Although the same characteristics generally occur throughout the region, each site may have unique features, determined by local climate, geology, and topography. On similar sites within this broader ecoregion, land managers can make similar conclusions about the land's sensitivity to acid rain, suitability for agriculture, or potential forest yield. Shirazi (1984) tested this idea in the subtropical mixed forests ecoregions of the southeastern United States and found that it enhanced the ability to predict potential environmental impacts of hazardous wastes. Thus, a map of the ecoregion can save land managers, governments, private land conservationists, and others considerable time and money.

Another example of this kind of use comes from the marine mountains ecoregion of western Canada. A Biogeoclimatic Ecosystem Classification System is used to define biologically equivalent site units based on characteristics of the ecosystem (Mackinnon et al. 1992). The system is now an integral part of silviculture in British Columbia. The treatment of particular sites for production forestry takes into account the understanding of the unit, based on experiences in dealing with equivalent units elsewhere in the region. The staff has compiled a body of management experiences about each biologically equivalent site unit in the region. The predicted responses of those units to particular management regimes are used when the staff decides on the most suitable crop tree species for reafforestation at a particular site. Therefore, the system is used to manage the sustained development of forest resources in western Canada.

Ecosystem management can greatly benefit from ecosystem classifications within a regional framework. Classification could help identify lands of similar attributes and facilitate data transfer from a site for which there is adequate data to other sites of the same type that lack similar data.

Enhancing Vegetation Maps

Resource management agencies like the USDA Forest Service use maps of vegetation for land-use decisions regarding fire management, habitat conservation, and watershed usage. Many of the maps, however, lack detail and are outdated. To create better vegetation maps, remote sensed data from satellite is being used in conjunction with geographic information systems.

Creating maps with spectral data from satellites is a relatively fast method for mapping vegetation. Problems can arise, however, when using satellite data alone. Many different vegetation types have the same **spectral signature**. In most vegetation maps, some different types get the same class label. However, in many cases, they should be managed differently. The result of one study (Brown et al. 1993) found ecoregions useful for resolving class confusion and correcting misclassifications involving a land-cover database for the conterminous United States. The most probable vegetation growing on a site within an ecoregion can be predicted from landform information if one knows the vegetation–landform relationships in various ecoregions.

All of the applications discussed in this chapter involve expanding our perspective to see the patterns that exist within a region. These patterns, interpreted in terms of process, can be very useful to land managers and others. In the next chapter, we summarize how various land-management agencies, as well as conservation organizations, use ecoregion maps.

CHAPTER 6

How Land-Management Agencies, Conservation Organizations, and Others Use Ecoregion Maps

M ost environmental concerns cross boundaries. Borders that separate countries, or the jurisdictions of regulatory agencies, are irrelevant to problems such as air pollution, declining anadromous fisheries, forest diseases, or threats to biodiversity. To address these problems, environmental planners and decision-makers must consider how geographically related systems are linked to form larger systems, like the ones discussed in this book. Issues that may appear to be local will often require solutions at the regional scale—working with the larger pattern, understanding how it works, and designing in harmony with it.

In 1976, I published a map of the ecoregions of the United States (Bailey 1976). As discussed below, the Forest Service adopted the map and classification upon which the map was based. Thus, for the purpose of this book, I refer to it as the Bailey–Forest Service classification. The classification was subsequently expanded to map North American and the world—continents as well as oceans. The maps have been updated periodically as new information and increased understanding were gained. My work stimulated other work and there have been numerous uses and applications of ecoregion classification systems for a broad range of purposes. For example, Brewer (1999) identified 350 plus citations to ecoregion classifications in the scientific literature, going back to 1976, when the first map was published. The Bailey–Forest Service classification was cited 145 times. Versions of the map appear in a number of textbooks [e.g., *Forest Ecology* by Barnes et al. (1998); *Ecology and Field Biology* by Smith and Smith (2001)] and atlases, including the 19th and 20th editions of *Goode's World Atlas* and the *Atlas of U.S. Environmental Issues* (Mason and Mattson 1990), as well as gardening books [e.g., *Landscaping with Nature* by Cox (1991)]. Ecoregions are used in the Peterson FlashGuides™ series on birds, butterflies, and trees. *Bird Watcher's Digest* publishes a backyard bird newsletter that uses the ecoregion format for gardening to attract wildlife (especially birds) and report on bird distribution. Microsoft® has incorporated the world map into its Encarta® Virtual Globe on CD-ROM. The U.S. Postal Service has begun issuing a series of stamps that feature 10 of the ecological regions in North America (Fig. 6.1). The series is called "Nature of America." One of the most unusual

Figure 6.1. United States postage stamps promoting the ecological regions of North America. Stamp design © 1999 U.S. Postal Service. Reproduced with permission. All rights reserved.

Table 6.1. Use of ecoregions classification based on publications, 1976–1998

Use	No. of citations	Percent
Avifauna	21	14
Ecoregion theory	20	14
Forestry	17	12
Climate and atmospheric studies	14	10
Remote sensing	13	9
Rangeland management	12	8
Ecoregion delineation	10	6
Aquatic fauna	9	6
Terrestrial fauna	9	6
Entomology	5	3
Water quality management	4	3
Wetlands	4	3
Geology	2	1
Lake water quality	1	1
Medical geography	1	1
Recreation management	1	1
Stream water quality and biological monitoring	1	1
Agricultural science	1	1
Total	145	100

Source: Data from Brewer (1999).

applications involves the geographic study of meteorites. By plotting meteorite incidents on ecoregion units, this new map is helping to explain the distribution of meteorite finds in North America. Other uses of the Bailey–Forest Service classification are listed in Table 6.1. Depending on the search engine being used on the World Wide Web, the term "ecoregion(s)" yields several hundred to several thousand results. This chapter will examine some of these ecoregion frameworks. It has been complied from a number of sources, the most important of which are listed in the Selected Bibliography section. Unless otherwise noted, the term "ecoregion(s)" refers to the Bailey–Forest Service framework.

Current Applications

There has been growing recognition of the need to address environment problems at the broader scale. As a result, land-management agencies and conservation organizations have expressed interest that ecoregion maps could play a role in meeting this need. Various alternatives have been suggested for mapping ecoregions and their component ecosystems. The National Director of Land Management Planning (now Ecosystem Management Coordination) of the Forest Service in Washington, DC asked me in the mid-1980s to synthesize these suggestions

USDA Frorest Service		
Bailey, 1988	Ecomap, 1993	
Ecoregion	Ecoregion	Domain Division Province
Landscape Mosaic	Subregion	Section Subsection
	Landscape	Landtype Assoc.
Site	Land Unit	Landtype Landtype Phase

Figure 6.2. Comparison of hierarchies used for ecological land classification by the U.S. Forest Service.

as well as the literature, and publish the results (Bailey 1985, 1987, 1988a). In 1992, a workshop held in Salt Lake City recommended that a Washington Office Task Team on ecological classification and mapping (Ecomap) standardize classification and mapping and establish a hierarchical structure of ecological units. This team adopted a hierarchy in part from my 1988 publication, *Ecogeographic Analysis: A Guide to the Ecological Division of Land for Resource Management* (Fig. 6.2). In 1993, as part of the National Hierarchical Framework of Ecological Units (Cleland et al. 1997), ecoregions were adopted by the Forest Service for use in ecosystem management. This framework is one of three federal major agency frameworks that are being used to develop a set of common ecological regions for the United States (McMahon et al. 2001). The intention is that the framework will foster an ecological understanding of the landscape, rather than an understanding based on a single resource, single discipline, or single agency perspective.

Ecological subregions also have been delineated and described for the whole country (Bailey et al. 1994; McNab and Avers 1994), the eastern United States (Keys et al. 1995), and some states, such as California (Goudey and Smith 1994), the three states comprising the Upper Great Lakes (Albert 1995), and some Forest Service administrative regions (e.g., Nesser et al. 1997). More detailed subdivisions—referred to as landtype associations and landtypes—have been developed for most National Forests and some states, such as Missouri and Minnesota, to assist with forest- and state-level analysis and planning.

Agency officials and scientists have noted that ecosystem management will require collecting and linking large volumes of scientific data about ecosystem structures, components, processes, and functions at several geographic scales to determine current conditions and trends. Currently, available data are often not compatible, and large gaps in information exist.

In addition to being noncompatible, data derived from inventories provides only knowledge of ecosystem components, not an under-

Figure 6.3. Ecoregional assessments completed, in progress, or planned by the U.S. Forest Service.

standing of structure and function. Designing for sustainability requires an understanding of how local ecosystems fit together and interact in a landscape and ecoregion (Chapter 1). Within an ecoregion, a repeated relationship, or pattern, occurs. This pattern reflects a process that is not discernable at the local ecosystem scale. Recognizing the formidable barrier posed by noncomparable and insufficient data, as well as too narrow a look at ecosystems, the Forest Service and its cooperators is changing its focus from local scale to ecoregional assessments of natural resources (Quigley et al. 1998). Figure 6.3 shows the location of these assessments. Other organizations, like the Conservation Biology Institute in Corvallis, Oregon, are conducting this kind of ecoregional assessment for the Pacific Northwest ecoregions, along the Pacific coast from Oregon to Alaska. The Wildlands Project in Tucson is doing the same thing to pursue long-term conservation in the Klamath–Siskiyou ecoregion that spans the California–Oregon border (Noss 1999/2000). Other efforts include assessments of the Southern Rocky Mountain ecoregion and the Grand Canyon ecoregion.

Roads have been identified as the major source of sediment in most forest watersheds due to surface erosion and landsliding. One of the most common forest road conditions leading to sedimentation of streams is where a forest road experiences erosion between cross-drains (culverts or open drains). The Cross-Drain (X-DRAIN) modeling software has been developed by Forest Service scientists to model potential sediment yield from roads as affected by climate, soil, and local topography. Climate data needed to run the model are derived from stations drawn from all of the dominant ecoregions in the country.[18] This is done to account for ecoregional differences in climate, thereby increasing the ability of the model to correctly predict sediment yield (Chapter 5). There are several applications for the X-DRAIN model. Optimum cross-drain spacing from existing or planned roads can be determined. Recommendations concerning road construction, realignment, closure, obliteration, or mitigation can be developed and supported by X-DRAIN results.

Traditionally, USDA Forest Service, Forest Inventory and Analysis (FIA) surveys typically are conducted to estimate and report information according to administrative divisions, such as counties, ownership class, and federal or state forest-management districts. This permits stakeholders to rapidly assess priorities and programs under their control or influence. A systematic sample of ground plots over the survey area is conducted to estimate the forest resource. However, surveys involving comprehensive sampling efforts will more accurately characterize unmonitored sites (plots) and discern relationships when samples are stratified according to ecologically similar areas such as ecoregions. O'Brien (1996) and Rudis (1998) found significant differences in the extent and conditions of forest resources among ecoregions in Utah and the southern United States, respectively. To facilitate integration of county-referenced information with areas of similar ecological potential, Rudis (1999) has assigned each county in the conterminous United States to the ecoregion framework.

In the development of databases, information from various inventories is starting to be classified by ecoregion. For example, ecoregion maps provide one of the geographic frameworks for the Forest Service's corporate database, referred to as Natural Resources Information System (NRIS), and the USDA Natural Resources Conservation Service's (NRCS) National Resources Inventory (NRI). Oak Ridge National Laboratory is collecting **net primary productivity** (NPP) data from sites around the world. The sites are plotted on a Bailey ecoregion map. The laboratory also has data from a global network of 150 micrometeorological tower sites that measure exchanges of carbon dioxide. The database is called FLUXNET, with sites referenced to ecoregions. Both databases can be accessed by visiting their Internet site, listed in the Resource Guide in this book. The climate network for USDA UVB (ultraviolet -B) Radiation Monitoring Program also plots their sites on an ecoregion map (see their Internet site, listed in the Resource Guide). This information is critical to the assessment of the potential impacts of increased ultraviolet radiation levels on agricultural crops and forests. Plots of available data for any date may be obtained by clicking on the map. Pictures of each climatological site will also appear. The International Institute for Applied Systems Analysis' (IIASA) Sustainable Boreal Forest Resources Project has developed an ecoregional database for Russia and Siberia.

Since 1976, I have elaborated the ecoregion concept and its application (Bailey 1983,1988a, 1989, 1991, 1995, 1996, 1998a, 1998b). My work stimulated other work on ecoregions—either expansion of the variety of applications of ecoregions or hybridization of my approach to ecosystem regionalization. It now appears that the influence of ecoregion mapping on research and planning efforts has been considerable. For example, the National Science Foundation is using ecoregions for its Long-Term Ecological Research (LTER) network to study global change. There are limits to the number of sites that can be established for monitoring changes in the global environment. Obviously, sites should be representative. Also, stations should be placed where they can detect change. The boundaries between ecoclimatic regions, or ecoregions, are potentially suitable for this purpose. At the boundaries occur the highest degree of instability of the ecosystems and greatest sensitivity of their components to various forms of pressure occurring.

In cases where establishing a new monitoring station is impractical, existing networks and individual studies have to be used. We can compare existing networks of ecosystem maps to see where representation is inadequate and where additional sites are needed. For example, the LTER sites are located in various ecoregions throughout the United States (Fig. 6.4). By relating the LTER sites to the map, we have a way to establish priorities for new sites. Furthermore, similar ecosystems

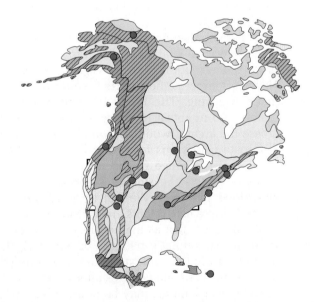

Figure 6.4. North American ecoregion boundaries and locations of Long-Term Ecological Research (LTER) sites. Compiled by LTER Network Office.

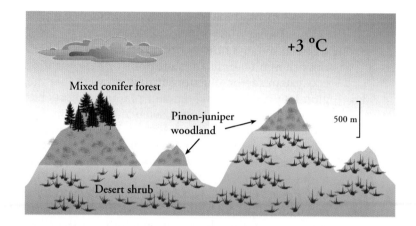

Figure 6.5. The approximate elevational boundaries of the vegetation types on the isolated mountain ranges of the Great Basin: *left,* today; *right,* in the future after a postulated climatic warming of approximately 3°C. Redrawn from *Macroecology* by James H. Brown. Copyright © 1995 The University of Chicago Press; reprinted with permission.

occur throughout each map unit. By comparing the location of the sites to the map, we can see how far the results of research at a particular site can be extended or transferred to analogous sites within an ecoregion.

Climate change may well affect the boundaries between ecoregions because we are dealing with a dynamic system. Figure 6.5 shows the predicted elevation shift of **vegetation zones** in the Great Basin in Nevada (temperate desert) that would occur assuming 3°C average climatic warming. The lower limit of woodland would shift approximately 500 m above its present elevation of 2280 m. This would decrease the area of woodland on all mountain ranges in the region and eliminate coniferous forest from some of them. Figure 6.5 also illustrates how a change in the larger system—in this case the lowland climate—affects the smaller systems (i.e., mountain ranges) that are embedded within it. We can think of these mountain ranges as "islands," where ecosystems are surrounded by an "ocean" of desert saltbush and sage. Just as the oceans control the continents through their influence on climatic patterns (Chapter 1), the mountain ecosystems operate within the context of the surrounding desert and are controlled by it. Should climate changes alter desert conditions, some ecosystems in the cooler habitats may be eliminated during droughts and hot periods. Also, desert pools in the Great Basin that are all that remain following the retreat of Pleistocene glaciers may dry up completely. This spells added hardships for the organisms that live there—the Devil's

Hole pupfish famous among them. Recolonization across deserts could be difficult or impossible.

With the changing climate, Arctic permafrost is melting, sea levels are rising, and areas affected by tropical diseases are expanding. Many species are already responding to warmer temperatures and encountering problems in the process. For example, Canada geese have begun arriving at their summer grounds weeks ahead of schedule, and taxing their food supply. Although the season also starts earlier at their food supply, this supply is timed by the length of day not temperature. As species' ranges change, land managers must start choosing sites that will provide refuges for species on the run. This will be difficult because they have to contend with considerable land fragmentation—the roads, cities, and farms that create obstacles in their paths. The message is clear. Not only must we design to sustaining present ecological systems, we must also design for the changes that are occurring.

Other organizations are starting to use this approach. The Nature Conservancy has shifted the focus from conservation of single species and small sites to conservation planning on an ecoregional basis (The Nature Conservancy 1997, Stolzenberg 1998). Figure 6.6 portrays "hotspots" of imperiled species by ecoregions. This view highlights the ecological importance of regions such as Appalachia and the mountains and deserts of the Southwest. The Nature Conservancy targets these hotspots for purchase of land. The Conservancy is also factoring climate change into their ecoregional and site-conservation plans. They are trying to predict how and where native plants and animals may be

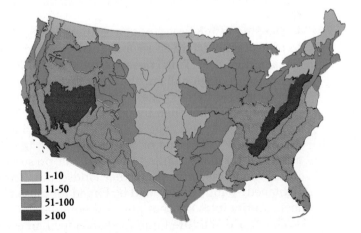

1-10
11-50
51-100
>100

Figure 6.6. Distribution of imperiled species by ecoregion. The Appalachians, the Great Basin, and coastal California stand out in this assessment of imperiled species. Reprinted from *Precious Heritage,* copyright © 2000 The Nature Conservancy and NatureServe; reprinted with permission.

most at risk from climate change and how they can best extend their protected lands to accommodate the migration of species.

The U.S. Fish and Wildlife Service adopted ecoregions years ago as the scientific framework for its National Wetlands Inventory (Cowardin et al. 1979), and the U.S. Geological Survey (USGS) uses ecoregions to assess gaps in the national network of biodiversity preserves (see, e.g., Stoms et al. 1998). The USGS also used ecoregions to design the National Trends Network to monitor the chemistry of wet deposition across the United States (Robertson and Wilson 1985). A network of 151 sites was proposed to provide, among other things, assurance that all areas of the country are represented in the network on the basis of regional ecological properties.

The National Park Service recently created a system of ecoregions to promote environmental sustainability. The Park Service's goal is to increase public education about ecosystems and ecoregional scale environmental conservation. Such ecoregion classifications are becoming increasingly common, and delineation techniques and criteria are becoming exceptionally diverse. For example, the Environmental Protection Agency (EPA) has adopted a national system of aquatic ecoregions developed by James Omernik (1987) to help water body managers address water quality concerns (Hughes 1997). The EPA approach to ecoregions has been applied to a number of states (e.g., Gallant et al. 1995) in order to refine the national system. The EPA is promoting ecoregional water resource planning. In Canada, an Ecoregions Working Group (1989) has been working to develop a map of ecoclimatic regions. Also, the Commission for Environmental Cooperation (1997) had delineated ecological regions for Mexico, Canada, and the United States.

The Sierra Club has created a "critical ecoregions" program designed to protect and restore 21 regional ecosystems in the United States and Canada (Fig. 6.7). This conservation organization has wholeheartedly embraced ecoregionalism, the idea that environmental problems are best addressed in the context of broad geographic areas defined by natural features rather than by political boundaries and borders. As Jane Elder (1994) points out in her "The Big Picture" article in the March/April 1994 issue of *Sierra* magazine, the Club has come to realize that there are limits to what it can preserve simply by drawing lines around visually spectacular landscapes. The Club feels that "enlightened" public policy will be forged for these regions, based on ecological principles and not political boundaries.

The World Wildlife Fund has developed an ecoregion classification system to assess the status of the world's wildlife and conserve the most biologically valuable ecoregions (Olson and Dinerstein 1998). The system is well established within the organization's structure and a world ecoregions map has recently been published in collaboration

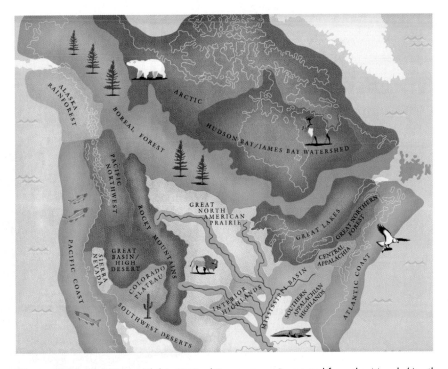

Figure 6.7. The Sierra Club's Critical Ecoregions. Reprinted from the March/April 1994 issue of *Sierra* magazine. Copyright © 1994 by the Sierra Club. Illustration by Max Seabaugh.

with the National Geographic Society. It was recently distributed to all U.S. schools.

Bat Conservation International of Austin, Texas is using the map of ecoregions in developing a regional priority matrix to gain a better understanding of the overall status of a given bat species. The purpose is to highlight the importance of a single region or multiple regions to the viability and conservation of each species. The matrix should also provide a means to prioritize and focus population monitoring, research, conservation, and the efficient use of limited funding and resources currently devoted to bats. They are also using ecoregions in their quest for better regional recommendations about bat-house designs and mounting recommendations to specific regions (Kiser and Kennedy 1997).

The Foreign Agricultural Organization (FAO) (1998) of the United Nations is developing a global ecological zoning system based on ecoregional concepts to collect information for the Forest Resource Assessment 2000. The FAO system follows Bailey (1989, 1995, 1998a) in using Köppen's climatic classification as a basis for the delineation of zones. Mapping will be carried out using **potential vegetation** maps

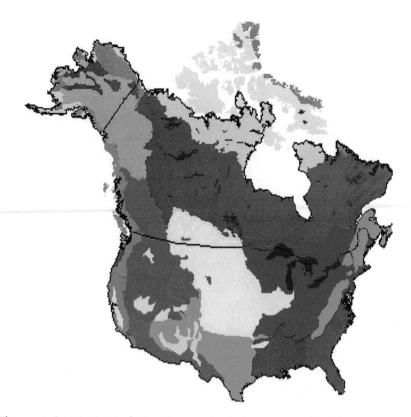

Figure 6.8. FAO's North American ecological zone map, level 3.

to define boundaries of climatic zones at three hierarchical levels. An example of third level mapping for North America is shown in Figure 6.8.

Realizing that traditional, commodity-focused research (e.g., potatoes) is inadequate, many agricultural researchers have moved toward a focus on research on land-use systems and ecoregions. This work is coordinated by the International Service for National Agricultural Research (ISNAR) (Bouma 1995). This organization defines an ecoregion as an area having a homogeneous biophysical environment, with comparable land-use systems and natural resource management problems, which allows research results to be applied to a large area. An ecoregional framework creates a platform for organizations from different agricultural sectors to work together, including those from forestry and fisheries.

The concept of an ecoregional approach to international agricultural research for sustainable agricultural production has been adopted by the Technical Advisory Group (TAG) of The Consultant Group on International Agricultural Research (CGIAR), which is part of FAO. Almost every CGIAR Centre has major ecoregional and interregional activities, including those in Sub-Saharan Africa, Asia, Latin America, West Africa and North Africa.[19] For instance, The International Potato Center (CIP), headquartered in Lima, Peru, has taken the lead in developing a global ecoregional research program for sustainable mountain agricultural development for the high Andes. This program is known as CONDESAN, Consortium for the Sustainable Development of the Andean Ecoregion. The International Institute of Tropical Agriculture (IITA) is shifting its traditional crop improvement and systems-based research strategy to an ecoregional research and development strategy. This holistic approach is expected to make IITA's research contributions more relevant to African farmers. Since 1989, the Centro Internacional de Mejoramiento de Maiz y Trigo (CIMMYT) has been collaborating in ecoregional initiatives, such as the Rice–Wheat Initiative for the Indo-Gangetic Plains and the Hillsides Initiative in Central America.

In cooperation with CGIAR, the International Food Policy Research Institute (IFPRI) has an initiative to develop a shared vision and consensus for action on how to meet future world needs by reducing poverty and protecting the environment. A key component of IFPRI's 2020 Vision initiative is a detailed assessment of future food needs, the potential for food and agricultural production to meet those needs, and the implications of agricultural activities for natural resources in major ecoregions of the developing world. They think ecoregional analysis is important not only because agricultural production potential differs significantly from one ecoregion to another, but also because the sustainable intensification of agriculture in the different ecoregions is likely to require quite different combinations of technology, policies, and institutional arrangements. The focus of one kind of analysis is malnutrition. Malnutrition is one of the most reliable indicators of endemic poverty. IFPRI has found that there are ecoregional dimensions to malnutrition. Analysis of ecoregions suggest that targeting of scarce resources to address the problems of poverty and malnutrition could potentially be improved by 2020 by considering linkages between ecoregional characteristics and poverty. Furthermore, ecoregional mapping helps to pinpoint areas prone to malnutrition.[20]

Recently, the Global Biodiversity Assessment of the United Nations Environment Programme (UNEP) concluded that the adverse effect of human impacts on biodiversity are increasing dramatically and threatening the very foundation of sustainable development. Realistically, only a relatively small portion of the total land area is likely to be devoted to biodiversity conservation; hence, it is critical to geographically identify areas rich in species diversity and endemism (species native or confined to a particular area) as a first step toward the protection of remaining natural habitat before they are destroyed. In the past, protected areas often have been set aside without regard to the biodiversity within their boundaries. As a result, many protected areas have little significance in terms of biodiversity and, conversely, many areas of habitat with significant biodiversity lack protection. The USGS in cooperation with UNEP and the National Aeronautic and Space Administration (NASA) is studying the relationship between biodiversity and ecoregions in Africa to determine if African ecoregions with significant biodiversity are adequately protected.[21]

Future Possible Applications

The U.S. Department of Defense is planning to use an ecoregional approach to evaluate military lands as to their capacity to sustain training and as to how well they are representative of global environments (Doe et al. 2000). The Center for Environmental Management of Military Lands (CEMML) at Colorado State University has researched and applied this approach. The locations of 31 major active Army installations were superimposed on a map delineating the boundaries of ecoregions (Fig. 6.9). In one study, ecoregions were classified as to their training resiliency or capability to support training while sustaining the existing ecological system. The sites with low resiliency were located in the deserts and semideserts of the West and Southwest. At some of these sites, tank tracks are still visible from training maneuvers conducted during World War II, which testifies to the long time it takes to recover from them. Other sites have higher resiliency and have potential to better sustain prolonged military land use. In a related study, CEMML researchers found the current Army land inventory adequate for conflict in temperate and dry ecoregions, particularly those which support a desert- or continental-type climate, but inadequate for areas represented by the savanna, rainforest, and Mediterranean ecoregions. A detail explanation of the application of ecoregions to military lands can be found on CEMML's website, listed in Appendix C.

Plant **germplasm** collections serve as important sources of genes for resistance to pests, diseases, and environmental stresses and help to ensure that potentially useful genetic variation is preserved. Identification and utilization of useful plant traits in germplasm collections can help to meet future demands for improved **cultivars**. To better utilize these genetic resources effectively, however, detailed knowledge about variations among individuals or groups of similar accessions is needed. Collections that have not been systematically characterized may also contain redundant accessions or lack representation of unique

Figure 6.9. Distribution of major Army installations in the United States, showing ecoregion boundaries, as delineated by Bailey. Compiled 1995 by the Center for Ecological Management of Military Lands (CEMML), Colorado State University, Fort Collins, Colorado.

or rare types. Accessions in germplasms collections often are not described by the ecoregion from which they were collected. Such information could be used to determine if other similar ecoregions exist where specific accessions may be adapted. Steiner and Poklemba (1994) found that the diversity of a common species of legume used for forage, birdsfoot trefoil, was highly correlated with the distribution of ecoregions as shown on the Ecoregions of the Continents map (Bailey 1989). The International Plant Genetic Resources Institute (IPGRI) has supported the creation and development of a number of ecoregional networks.

Ecoregions could be used to extrapolate data from Long-Term Ecological Research sites (LTERs). The system of LTERs was created to promote baseline studies of ecological conditions, but difficulties have arisen with data uniqueness. One of the difficulties with studies based on data obtained from LTERs is that data are only useful for the specific area in which the research was conducted. Ecoregions could be used as extrapolation units for the information discovered in LTERs. The information could then be used for resource management applications, rather than solely for baseline research.

Anthropogenic and climatic change could yield ecoregions that are much different, or less useful, after many years. For example, climate change could influence ecoregional boundaries and new ecoregions could be created by acid deposition as a result of industrial activity. Therefore, temporal variability is an important research issue. Brewer (1999) has suggested the addition of an expiration date to ecoregion classifications. With the increased ease of computer mapping and on-line distribution, ecoregion developers could update versions of the maps in a short period, thus keeping the framework current.

The rapid expansion of urban areas is also an area of concern that Brewer (1999) has identified. Large cities such as Las Vegas and Phoenix are significantly changing the environmental surroundings of the region. Increased moisture from lawns and golf courses has caused microclimatic change and the smog from automobile emissions affects precipitation and temperature patterns. Soils are also altered with chemical and organic inputs, further altering the natural status of the region. Roads and parking lots prevent native plant species from establishing themselves, allowing non-native flower and garden varieties to spread throughout the region. As impervious surfaces, roads and parking lots generate huge amounts of stormwater containing a host of automobile contaminants—oil, rubber, organic compounds, and many other residues. Today, the quantity of petroleum residue washed off these surfaces each year exceeds the total spillage from oil tankers and barges worldwide. Between 1950 and 1980, highway salt applications in the United States increased 12 times. The salt content in expressway drains during spring runoff may be 100 times greater than natural levels in freshwater. These alterations might create the need for the delineation of a new class of urban ecoregions.

According to University of Missouri rural sociologist Elizabeth Barham, rural communities, and particularly communities heavily dependent on food and agriculture-related production, have often found themselves at a disadvantage in the emerging world of free trade.[22] In many instances, localities under economic pressure from international competition have developed labels of origin for their products in an attempt to enlist the loyalties of the consumer and attract him or her to the image associated with a particular place. Such labels offer customers the choice of purchasing local products to support the local economy, for example, or buying imported products that are associated with particular values such as fair labor practices, environmental-friendly production (e.g., maintains wildlife corridors, avoids clearing virgin rainforest), or maintenance of specific cultural tradition linked to a particular place. Starbucks, for example, states their commitment to origin, or place, as follows:

> "Contributing positively to our communities and our environment" is a critical guiding principle in Starbucks Mission Statement and is the foundation from our involvement in coffee origin countries. We're committed to addressing social and environmental issues in order to help sustain the people and places that produce Starbucks coffees.[23]

These labels encompass the way in which the product itself fits into the overall environment. As such, the term "eco-labelling" is sometimes applied. The popularity of these locally embedded food systems has continued to rise among supporters of sustainable agriculture. Working with groups of people who are trying to label their production based on place provides one way for introducing ecoregion sensitivity into local production methods.

This idea of developing labels tied to location of local food production is similar to the concept of the French "terroir." It is a commonly held belief in France as well as across the European continent that the taste of a particular food will vary from region to region, reflecting the physical and cultural constituents of the countryside. For years, "terroir" has meant a commitment to ingredients locally produced by small, independent farmers and cooperatives. The label-of-origin system used in France originally evolved to protect its wine regions (Wilson 1999).

The relationship between vegetation and landform position changes from region to region, reflecting the effect of the macroclimate. A species occupies different positions in the landscape relief, for example, when proceeding from north to south. The result is that, for example, in the northern temperate climates, a species may prefer north-facing slopes in southern parts of its range and, conversely, south-facing slopes in northern parts of its range. Knowledge of these relationships is important for extending results of research and management experience, and for designing sampling networks for inventory and monitoring. These relationships have been extensively studied in some regions (e.g., southern Appalachian Mountains) but, unfortunately, not in others. In fact, most present inventories provide only a knowledge of individual ecosystem components and not an understanding of how the components fit together or relate to one another. Future inventories could be designed to capture these relationships.

We are surrounded by information—more than we know how to comprehend; or as E.O. Wilson (1998) writes in his book *Consilience: The Unity of Knowledge,* "We are drowning in information while starving for wisdom." There is a saying "if you don't know what to do; build a database." Databases abound but with little synthesis of the infor-

Summary and Conclusions

As an integral part of systems, the impact of humans is a key factor in sustainability and must be studied on a large scale. This book summarizes the ideas about how an understanding of ecoregions can contribute to land-management and conservation programs through the integration of ecoregional mapping and ecological design. Ecological design considers the patterns of processes that shape a region, including the climate, topography, soils, vegetation, fauna, and culture, and incorporates these factors in design and planning. Conclusions drawn from these ideas are discussed and illustrated in the following four sections.

Spatial Pattern Matters

Ecosystems do not exist in isolation. Spatial pattern matters. The juxtaposition of local sites is important to how they behave. Most studies focus on site-specific impact for individual ecosystems: for example, a gully (Fig. 7.1) formed in response to clearing the forest in the Lake Tahoe basin for housing development before land capability constraints went into effect. It is no longer appropriate to plan based on totals or averages of resource outputs over an area or water flows from a watershed. Rather, the arrangement of land use and buildings is crucial to planning, conservation, management, and sustainable design.

Figure 7.1. Forests have been removed in the Lake Tahoe basin for housing development. Removal of vegetation has exposed soil to greatly accelerated erosion. In addition, the streets intercept and concentrate runoff, greatly increasing its erosive force. Unknown photographer.

Context Is Usually More Important Than Content

Furthermore, context is usually more important than content. Usually, a place is described by its characteristics. Yet, as Dranstad et al. (1996) observed, the characteristics of the surrounding adjacent areas and land uses, of the upstream–upwind–upslope, and of the downstream–downwind–downslope areas are usually more important descriptors of the location. The process of expanding our perspective sheds new light on the interrelationships among ecosystems so that we will be better able to predict the impacts of human activity and to design accordingly within the larger spatial context. Thinking ecoregionally makes us aware of these interrelationships. The effects of land development in the Lake Tahoe basin (Fig. 7.1) could have been predicted and prevented if the climate and processes of the ecoregion where first taken into consideration. In regions like the Sierran mountains of California with a summer-dry subtopical climate, the granitic bedrock disintegrates to individual grains and is highly susceptible to accelerated erosion, particularly when the vegetation is removed. What happened in

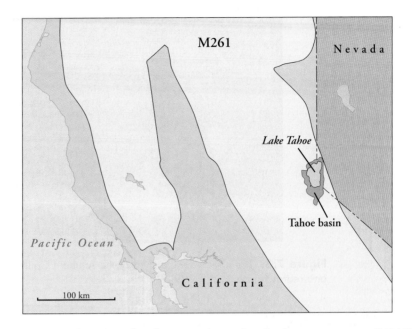

Figure 7.2. The Lake Tahoe basin site lies within the Sierran ecoregion (M261), as mapped by Bailey (1995).

the Lake Tahoe basin was not isolated to the basin. It was symptomatic of problems and processes over the ecoregion as a whole that extends from the mountains of the Pacific Coast of Northern California eastward to Nevada and from Oregon down south to the southern border of the Sierra Nevada Mountains (Fig. 7.2). Analyzing natural systems at the site must also take into consideration the ecoregional patterns and processes.

That the character of landscapes with identical geology will differ significantly between different climatic regions has already been demonstrated (Chapter 3). Mountainous areas exhibit an altitudinal zonation of climate: The forest zone and snow line are found at completely different heights, and in each zone, the individual landform elements take on climatically controlled characteristics. Ecoregions of different climates differ significantly. Thus, the ecoregional patterns and processes of a mountainous region in a dry-summer subtropical climate, like the Sierran mountains, will differ significantly from a mountainous region in, say, the humid tropics. Recognition of processes and regional context is growing quickly in the field of design. The design of sustainable environments will be successful to the degree that they are sympathetic to ecoregional differences.

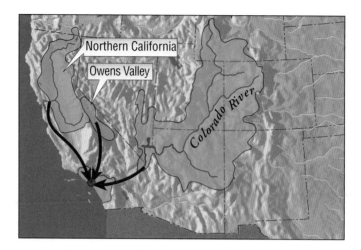

Figure 7.3. The effective watershed of the southern California region. Modified and redrawn from Lyle (1999).

Matching Development to the Limits of the Regions Where We Live

Our ecological crises have resulted, in part, from a failure to match human development to the limits of the regions where we live. We have overstepped these limits. A good example is provided by vast water diversion projects that maintain life in desert towns like Las Vegas, Phoenix, and Los Angeles (Fig. 7.3). Water is supplied to southern California from three distant watersheds through three different aqueducts having a total length of over 1290 km. Competing uses and ecological concerns in all three watersheds are increasingly threatening to limit supplies. Proposals have been made to transport water from as far away as the Yukon in the Subarctic ecoregion to alleviate the limited supply. An alternative would be to design for efficient use and reuse of water.

Throughout history, overstepping the limits of natural resources has lead to conflict, even war, throughout the world. This is particularly true where two or more states divide a single ecoregion (Byers 1991). In these cases, conflict may arise over the apportioning of the natural resources of the region or over transboundary pollution.

What Is the Next Step?

In contrast, by matching development and use of the landscape to its inherent geometry, we allow ecological patterns to work *for us*. We can use natural drainage instead of storm drains, wetlands instead of sewage-treatment plants, and indigenous materials rather than imported ones. Although a sense of place is usually thought to be governed by nature, our modern human landscapes have developed a potency which sometimes sets them apart. Take, for instance, the quilted undulating landscapes of Ireland, where the fields have been divided by dry stonewalls. The builders have fitted the stones together so skillfully that they interlock without the need for mortar, and some walls have been standing for centuries. They faithfully reflect the underlying geology, because it costs too much to haul stone any great distance for anything as common as a farm wall, and most of the stone is quarried on the premises. The color of the stone echoes the color of the bedrock to produce a natural flavor in the landscape. Another example are the adobe buildings in New Mexico, which seem to emerge from the earth itself (Fig. 7.4). These classic vernacular buildings are

Figure 7.4. Pam Furumo, *House at Arroyo Hondo*, watercolor, 1987. This house near Taos, New Mexico, is set in the matrix of the native landscape, ensuring continuity and creating a definite sense of place. Reproduced with permission.

a unique regional style in the sense that they mimic the characteristics of the environment and of the ecological region.

The course is clear: To conserve resources, we must be attentive to the sense of place, and match development to the limits of regions where we live. To accomplish this, we must consider the regional patterns and processes that shape a region and design land use and buildings accordingly.

Most of us live in a built environment where ecological patterns are hidden from our everyday awareness. Many of us work in modern buildings where often we cannot even open the windows. Van der Ryn and Cowan (1996, p. 161) ask, "What do we learn from this kind of 'nowhere' environments?" Well, it is not surprising that in these denatured places that are heavily dependent on technology to maintain, we lose some sensitivity to the natural world that sustains us. What we need is more of a connection to regions in which we live and the processes that are endemic to it. Through continued education, policy changes, public awareness, and application of the principles of ecoregional design, we *can* achieve this objective, and a more sustainable future.

Ecological Climate Zones

Köppen group and types	Ecoregion equivalents
A Tropical and humid climates	Humid tropical domain (400)
Tropical wet (Ar)	Rainforest division (420)
Tropical wet–dry (Aw)	Savanna division (410)
B Dry climates	Dry domain (300)
Tropical/subtropical semiarid (BSh)	Tropical/subtropical steppe division (310)
Tropical/subtropical arid (BWh)	Tropical/subtropical desert division (320)
Temperate semiarid (BSk)	Temperate steppe division (330)
Temperate arid (BWk)	Temperate desert division (340)
C Subtropical climates	Humid temperate domain (200)
Subtropical dry summer (Cs)	Mediterranean division (260)
Humid subtropical (Cf)	Subtropical division (230)
	Prairie division (250)[a]
D Temperate climates	
Temperate oceanic (Do)	Marine division (240)
Temperate continental, warm summer (Dca)	Hot continental division (220)
Temperate continental, cool summer (Dcb)	Prairie division (250)[a]
	Warm continental division (210)
	Prairie division (250)[a]
E Boreal climates	Polar domain (100)
Subarctic (E)	Subarctic division (130)
F Polar climates	
Tundra (Ft)	Tundra division (120)
Ice Cap (Fi)	Icecap division (110)

Definitions and Boundaries of the Köppen–Trewartha System

Ar	All months above 18°C and no dry season
Aw	Same as Ar, but with 2 months dry in winter
BSh	Potential evaporation exceeds precipitation, and all months above 0°C
BWh	One-half the precipitation of BSh, and all months above 0°C
BSk	Same as BSh, but with at least 1 month below 0°C
BWk	Same as BWh, but with at least 1 month below 0°C
Cs	8 months 10°C, coldest month below 18°C, and summer dry
Cf	Same as Cs, but no dry season

Do   4–7 months above 10°C, coldest month above 0°C
Dca   4–7 months above 10°C, coldest month below 0°C, and warmest month above 22°C
Dcb   Same as Dca, but warmest month below 22°C
E   Up to 3 months above 10°C
Ft   All months below 10°C
Fi   All months below 0°C

A/C boundary = equatorial limits of frost; in marine locations, the isotherm of 18°C for coolest month
C/D boundary = 8 months 10°C
D/E boundary = 4 months 10°C
E/F boundary = 10°C for warmest month
B/A, B/C, B/D, B/E boundary = potential evaporation equals precipitation.
[a]Köppen did not recognize the Prairie as a distinct climatic type. The ecoregion classification system represents it at the arid sides of the Cf, Dca, and Dcb types, following Borchert (1950).
Source: Based on the Köppen (1931) system of classification, as modified by Trewartha (1968).

Climate Diagrams

Climate diagrams of representative climate stations (based on Walter and Lieth 1960–1967, Walter et al 1975).

Polar Domain

110 Ice Cap

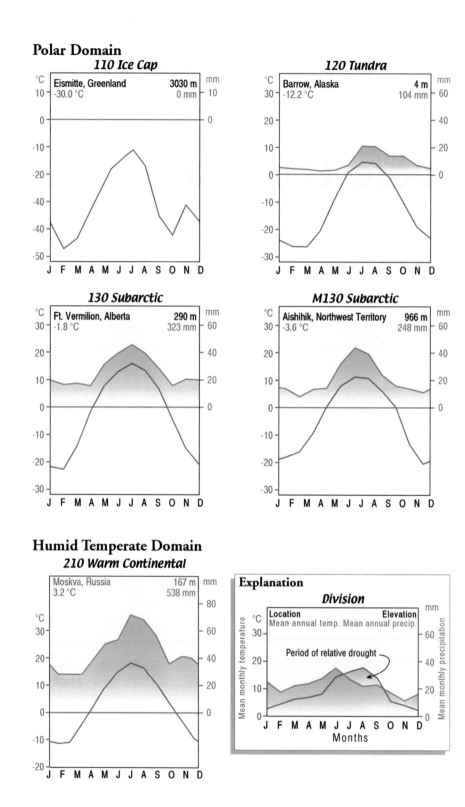

Eismitte, Greenland — 3030 m
-30.0 °C — 0 mm

120 Tundra

Barrow, Alaska — 4 m
-12.2 °C — 104 mm

130 Subarctic

Ft. Vermilion, Alberta — 290 m
-1.8 °C — 323 mm

M130 Subarctic

Aishihik, Northwest Territory — 966 m
-3.6 °C — 248 mm

Humid Temperate Domain

210 Warm Continental

Moskva, Russia — 167 m
3.2 °C — 538 mm

Explanation

Division

Location — Elevation
Mean annual temp. Mean annual precip.

Period of relative drought

Mean monthly temperature

Mean monthly precipitation

Months

Humid Temperate Domain

220 Hot Continental

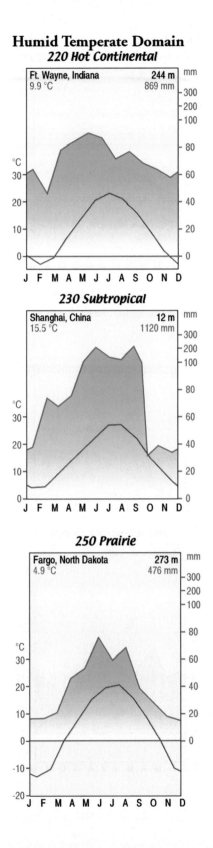

Ft. Wayne, Indiana — 244 m
9.9 °C — 869 mm

230 Subtropical

Shanghai, China — 12 m
15.5 °C — 1120 mm

250 Prairie

Fargo, North Dakota — 273 m
4.9 °C — 476 mm

M220 Hot Continental

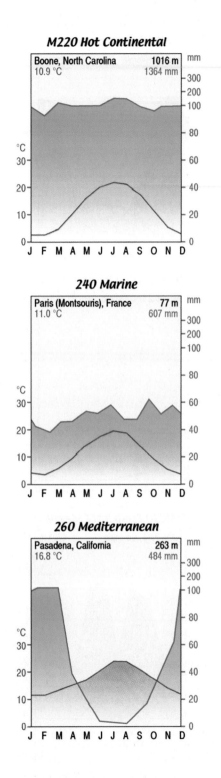

Boone, North Carolina — 1016 m
10.9 °C — 1364 mm

240 Marine

Paris (Montsouris), France — 77 m
11.0 °C — 607 mm

260 Mediterranean

Pasadena, California — 263 m
16.8 °C — 484 mm

M260 Mediterranean

310 Tropical/Subtropical Steppe

DryDomain

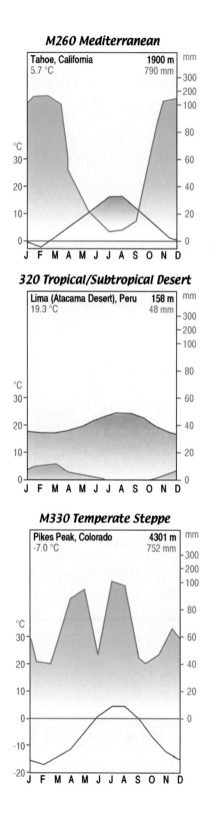

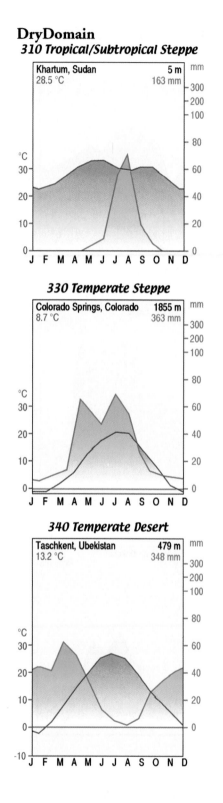

320 Tropical/Subtropical Desert

330 Temperate Steppe

M330 Temperate Steppe

340 Temperate Desert

Tahoe, California   1900 m
5.7 °C   790 mm

Khartum, Sudan   5 m
28.5 °C   163 mm

Lima (Atacama Desert), Peru   158 m
19.3 °C   48 mm

Colorado Springs, Colorado   1855 m
8.7 °C   363 mm

Pikes Peak, Colorado   4301 m
-7.0 °C   752 mm

Taschkent, Ubekistan   479 m
13.2 °C   348 mm

Humid Tropical Domain

410 Savvanna

M410 Savanna

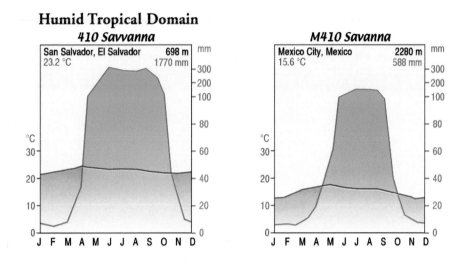

San Salvador, El Salvador   698 m
23.2 °C   1770 mm

Mexico City, Mexico   2280 m
15.6 °C   588 mm

420 Rainforest

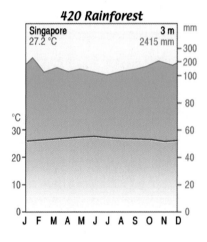

Singapore   3 m
27.2 °C   2415 mm

Resource Guide

This list is for information purposes only. Inclusion does not imply endorsement, nor is criticism implied of similar resources not mentioned.

Organizations, periodicals, and other materials on sustainable design and ecoregions briefly mentioned in the text are listed below. Books and journal articles are listed separately in the Selected Bibliography.

Ecoregions of the World. 1998. *In:* Microsoft® Encarta® Virtual Globe CD-ROM. 1998 ed. Redmond, WA: Microsoft.

Rheingold, H. 1994. *The Millennium Whole Earth Catalog.* San Francisco: Harper. 384 pp. This catalog is an evaluation and access tool. Their reviewers introduce books, magazines, tools, software, videotapes and audiotapes, organizations, services, and ideas. There are numerous listings on sustainability. The listings are continually revised and updated by users and staff. The latest news and access is published in the bimonthly magazine *Whole Earth.*

World ecoregions, types of natural landscapes. 2000. *In*: Hudson J.C; and Espenshade E.B. (eds.). *Goodes World Atlas*, 20th edn. Chicago: Rand McNally: 28–29. Scale = 1:77,000,000.

Peck, B. 2000. *Guide to North American Meteorites* [map]. Scale = 1:5,000,000. meteoritemaps.com. Meteorites finds are plotted on a map showing ecoregion boundaries. Even though meteorites fall randomly across the globe, they are subjected to terrestrial forces that vary greatly with ecoregion. Thus, meteorite finds are not evenly distributed.

Peterson FlashGuides™ Series 1996: *Backyard Birds* by Roger Tory Peterson, edited with text by Kevin J. Cook; *Butterflies* by Paul Opler and Amy Bartlett Wright; *Trees* by George A. Tetrides, Olivia Petrides, and Janet Wehr. New York: Houghton Mifflin. Organized according to the U.S. Forest Service's ecoregion for the United States and Canada.

Federal Information Sources

USDA Forest Service
Ecoregion Studies Program at the
Inventory & Monitoring Institute
2150 Centre Avenue, Bldg.A
Fort Collins, CO 80526
(970) 295-5740
Guidance and resources for ecoregional-based planning and design.

Nonprofit and Other Organizations

Center for Maximum Potential Building Systems
86604 F.M. 969
Austin, TX 78724
(512) 928-4786
"Max's Pot" has long been a leader in alternative building systems and materials research, the use of local materials, and advanced energy and wastewater system for buildings.

International Institute for Bau-Biologies™ & Ecology, Inc.
P.O. Box 387
Clearwater, FL 33757
(813) 461-4371
Offers information on "healthy homes," including home-study courses, consulting, seminars, lectures, and design.

Land Institute, The
2440 East Waterwell Road
Salina, KS 67401
(913) 823-5376
Wes Jackson and colleagues perform innovative research intended to create a sustainable prairie agriculture based on native perennial species.

Nature Conservancy, The
4245 North Fairfax Drive
Suite 100
Arlington, VA 22203
(703) 841-5300
http://www.tnc.org
The mission of The Nature Conservancy is to preserve the plants, animals, and natural communities that represent the diversity of life on Earth by protecting the lands and waters they need to survive. They work to accomplish this mission through ecoregion-based conservation. Their journal *Nature Conservancy* is published bimonthly.

Planet Drum Foundation
P.O. Box 3121
San Francisco, CA 94131
(415) 285-6556
http://www.planetdrum.org
Planet Drum was founded in 1973 to provide an effective grassroots approach to ecology that emphasizes sustainability, community self-determination, and regional self-reliance. In association with community activists and ecologists, Planet Drum developed the concept of bioregion: a distinct area with coherent and interconnected plant and animal communities, and natural systems, often defined by a watershed. Activities include publishing, lectures, seminars, workshops, and networking.

Rocky Mountain Institute
1739 Snowmass Creek Road
Snowmass, CO 81654
(303) 927-3128
Conducts research and outreach programs to foster the efficient and sustainable use of resources.

Wildlands Project, The
1955 W. Grant Road
Suite 145
Tucson, AZ 85745
(520) 884-0875
http://www.twp.org
This project is working on a long-term biodiversity and wilderness recovery plan for North America. The quarterly journal *Wild Earth* is the publishing wing of the project.

Periodicals

American Bungalow
123 S. Baldwin Avenue
Sierra Madre, CA 91025
(626) 355-1651
Published four times a year. Features include articles on rehabbing a historic bungalow, mission furniture, bungalow gardening, Art and Crafts collectibles, and plans for new construction. Book reviews.

Natural Home
P.O. Box 552
Mt. Morris, IL 61054
(800) 340-5846
Published bimonthly by Interweave Press, this magazine publishes articles on earth-inspired living, including sustainable interior design, building, landscaping, and cooking. A list of native plant experts for each ecoregion is listed in the March/April 2001 issue on p. 76.

Places, A Forum of Environmental Design
P.O. Box 1897
Lawrence, KS 66044
(800) 627-0629
Published three times a year. A forum in which designers, public officials, scholars and citizens can discuss issues vital to environmental design, with particular emphasis on public spaces in the service of the shared ideals of society.

Plateau Journal
Museum of Northern Arizona
3101 N. Fort Valley Road
Flagstaff, AZ 86001
(520) 774-5211, ext. 273
A semiannual magazine dedicated to the land and peoples of the Colorado Plateau ecoregion.

Wild Earth
P.O. Box 455
Richmond, VT 05477
(802) 434-4077
Published quarterly by The Wildlands Project. Publishes articles that meld conservation biology and wilderness activism. Book reviews.

Whole Earth
P.O. Box 3000
Denville, NJ 07834
(888) 732-6739
Published quarterly. This is the magazine of the *Whole Earth Catalog.* Publishes reviews of tools, ideas, and practices. As the name suggests, its scope encompasses the whole Earth. Book reviews.

Wildflower
4981 Highway 7 East
Unit 12A, Suite 228
Markham, Ontario, Canada L3R 1N1
Published quarterly by the Canadian Wildflower Society, this black-and-white magazine carries informative profiles of native plants.

Wild Garden
Unfortunately, this magazine ceased publication in 2000. The premier issue of 1998 contains a native plant resources section with the following information: Wild Gardens You Can Visit, Resources for Native Plants and Seeds, Organizations & Associations, Native Landscape Architects and Designers, Resources in Print.

Landscape and Urban Planning: An International Journal of Landscape Ecology, Landscape Planning, and Landscape Design
Elsevier Science
633 Avenue of the Americas
New York, NY 10010
(212) 633-3730

On the Internet

USDA Sustainable Agriculture Network
http://www.sare.org
Information about the Sustainable Agriculture Research and Education program of the Department of Agriculture.

USDA Forest Service, Ecoregion Studies Program at the Inventory & Monitoring Institute
http://www.fs.fed.us/institute/ecolink
Guidance and resources for ecoregional-based planning and design. Download ecoregion maps.

USDA Forest Service, Sustainability Home Page of Northeastern Area
http://www.na.fs.fed.us/sustainability
Loads of information on sustainability assessments, forest resource
planning, and ecological information and planning.

The National Park Service, Sustainable Information Directory
http://www.nps.gov/sustain
A clearinghouse for resources on sustainability.

Sierra Club Critical Ecoregions Program
http://www.sierraclub.org/ecoregions
Map and information about 21 critical ecoregions in North America.

Sierra Club Sprawl Page
http://www.sierraclub.org/sprawl
Access reports and activist resources on sprawl, smart growth, transportation, and livable communities

World Wildlife Fund Terrestrial Ecoregions of the World
http://www.worldwildlife.org/wildworld
Information and images for land-based ecoregions on the planet.

USDA UVB (ultraviolet–B) Radiation Monitoring Program
http://uvb.nrel.colostate.edu/UVB/uvb_climate_network.html
Climatological sites located on ecoregion map of the United States.

Oak Ridge National Laboratory Net Primary Productivity (NPP) Database
http://www-eosdis.ornl.gov/NPP/html_docs/na_npp_site.html
Database sites located on ecoregion map of North America and globe.

Center for Environmental Management of Military Lands (CEMML)
http://www.cemml.colostate.edu/bailey_ecoregion.htm
Applications of Bailey's ecoregions to military lands.

National Geographic Society
http://www.nationalgeographic.org
In conjuction with their July 2001 issue *National Geographic* lets you
virturally explore a smart growth suburb that depicts new urbanist
ideas for fighting sprawl: mixed-use zoning, pedestrian-friendly streets,
transit, town centers.

Colorado Sustainability Project
http://www.sustainablecolorado.org
Colorado's sustainable development information clearinghouse.

Congress for New Urbanism
http://www.cnu.org
Learn about the philosophy behind New Urbanism.

American Farmland Trust
http://www.farmland.org
Learn about efforts around the nation to preserve farmland and promote environmentally responsible farming practices.

Ecoregional Planning—The Nature Conservancy
http://www.consci.org/ERP/EcoregionalPlanning.cfm
Information on ecoregional planning and several links to various resources and maps such as ecoregional plans status.

Center of Excellence for Sustainable Development
http://www.sustainable.doe.gov
A project of the U.S. Department of Energy. A plethora of information on green buildings, green development, land-use planning, ecological landscape planning, transportation, sprawl, smart growth, top websites on sustainable community development, and breaking news.

Videos

Considering All Things: Healthy, Productive Ecosystems
This 19-min video explores the concept of ecosystem scale from the microscopic to the global. We travel from the forest soil to the forest canopy and beyond, finally circling the globe to join in an important wildlife conservation project. More information on this 1996 program can be obtained from USDA Forest Service, Office of Communications, P.O. Box 96090, Washington, DC 20090, (202) 205-1760.

Subdivide and Conquer: A Modern Western
Examines the causes of sprawl and its effect on our communities and the environment and suggests remedies along with examples of sound public policy and good land-use planning. Available from Bullfrog Films, P.O. Box 149, Oley, PA 19547, (800) 543-3764.

Common and Scientific Names

Plants

Aspen, quaking	*Populus tremuloides*
Beech	*Fagus* spp.
Birch, paper	*Betula papyrifera*
Birdsfoot trefoil	*Lotus corniculatus*
Bluegrass	*Poa pratensis*
Buffalo grass	*Buchloe dactyloides*
Cactus, jumping cholla	*Opuntia fulgida*
Cactus, prickly pear	*Opuntia phaeacantha*
Cactus, saguaro	*Cereus giganteus*
Cattail, common	*Typha latifolia*
Cedar, northern white	*Thuja occidentalis*
Chokecherry, western	*Prunus virginiana* subsp. *Melanocarpa*
Cottonwood	*Populus deltoids*
Creosote bush	*Larrea tridentata*
Cypress	*Taxodium distichum*
Dogwood, mountain	*Cornus nuttalli*
Eucalyptus	*Eucalyptus* spp.
Fir, balsam	*Abies balsamea*
Fir, Douglas	*Pseudotsuga menziesii*
Fir, white	*Abies concolor*
Grama, blue	*Bouteloua gracilis*
Hackberry	*Celtis occidentalis*
Hickory	*Carya ovata*
Ironwood	*Ostrya virginiana*
Ironwood, desert	*Olneya tesota*
Juniper, Rocky Mountain	*Juniperus scopulorum*

Juniper, Sierra	*Juniperus occidentalis*
Magnolia, southern	*Magnolia grandiflora*
Magnolia, sweet bay	*Magnolia virginiana*
Mahogany	*Swietenia mahogami*
Maple, big tooth	*Acer saccharum* susp. *Grandidentatum*
Maple, Rocky Mountain	*Acer glabrum*
Maple, sugar	*Acer saccharum*
Mesquite	*Prosopis* spp.
Oak, bur	*Quercus macrocarpa*
Oak, California black	*Quercus celloggii*
Oak, California scrub	*Quercus dumosa*
Oak, Gambel's	*Quercus gambelii*
Oak, interior live	*Quercus wislizenii*
Oak, red	*Quercus rubra*
Ocotillo	*Fouquieria splendens*
Olive, Russian	*Elaeagnus angustifolia*
Palo verde, bule	*Cercidium floridum*
Pine, jack	*Pinus banksiana*
Pine, ponderosa	*Pinus ponderosa*
Pine, white	*Pinus strobus*
Plum, pigeon	*Coccoloba diversifolia*
Redwood	*Sequoia sempervirens*
Sagebrush	*Artemisia* spp.
Saltbush	*Atriplex corrugata*
Sourwood	*Oxydendrum arboreum*
Spruce, black	*Pinus mariana*
Spruce, Norway	*Picea abies*
Spruce, white	*Picea glauca*
Sycamore, Arizona	*Platanus wrightii*
Tamarisk	*Tamarix ramosissima*
Thistle	*Cirsium* spp.
Tupelo	*Nyssa aquatica*
Twinberry	*Myrcianthes fragans*
Yew, western	*Taxus brevifolia*

Animals

Bat	Numerous genera and species
Bear, grizzly	*Ursus horribilis*
Bison, American	*Bison bison*
Coyote	*Canis latrans*
Elk, American	*Cervus canadensis*
Goose, Canada	*Branta canadensis*

Hawk, Swainson's	*Buteo swainsoni*
Hog sucker, northern	*Hypentelium nigricans*
Jackrabbit	*Lepus* spp.
Lion, mountain (puma)	*Felix concolor*
Locust (grasshopper)	Numerous genera and species
Phainopepla	*Phainopepla nitens*
Pigeon, passenger	*Ectopistes migratorius*
Prairie dog	*Cynomys* spp.
Pronghorn (antelope)	*Antilocapra Americana*
Pupfish, Devil's Hole	*Cyprinodon diabolis*
Robin, American	*Turdus migratorius*
Salmon	*Oncorynchus*
Sheep, bighorn	*Ovis canadensis*
Thrasher, brown	*Toxostoma rufum*
Waxwing, bohemian	*Bombycilla garrulus*
Wolf, gray	*Canis lupus*

Conversion Factors

For readers who wish to convert measurements from the metric system of units to the inch–pound–Fahrenheit system, the conversion factors are as follows.

Multiply	By	To obtain
Millimeters	0.039	Inches
Centimeters	0.394	Inches
Meters	3.281	Feet
Kilometers	0.621	Miles
Square meters	10.764	Square feet
Square kilometers	0.386	Square miles
Hectares	2.471	Acres
Celsius	1.8 + 32	Fahrenheit
Hectare-meters	8.107	Acre-feet

Glossary of Terms as Used in This Book

Alfisol   soil order consisting of soils of humid and subhumid climates, with high base status and argillic horizon.

Anadromous fisheries   migrating from saltwater to spawn in freshwater, such as salmon.

Aquifer   a body of rock that conducts groundwater in usable amounts.

Arcade   covered walkway at the edge of a building.

Arid   dry, with limited vegetation, rainfall less than about 250 mm (10 in.) and great excess of evaporation over precipitation.

Aridisol   soil order consisting of soils of dry climates, with or without argillic horizons, and with accumulations of carbonates or soluble salts.

Arroyo   in southwest United States, steep-sided dry valley, usually inset in alluvium.

Aspect   see Exposure.

Base level   a hypothetical level below which a stream cannot erode its valley, and thus the lowest level for denudation processes.

Basin   see *Watershed*.

Biodiversity   variety of life and its processes, including the variety in genes, species, ecosystems, and the ecological processes that connect everything in ecosystems.

Biogeographical region   one of eight continent-sized or subcontinent-sized areas of the biosphere, each representing of evolutionary core areas of related fauna (animals) and flora (plants); for example, the Neotropical of Wallace (1876).

Biogeoclimatic   used in association with classification of ecosystems and expressed as types of vegetation, climate and site characteristics. (The combination of biological, geological, and climatic factors affecting distribution patterns.)

Biome   a geographical region which is classified on the basis of the dominant or major type of vegetation and the main climate; for example, the temperate biome is the geographical area with a temperate climate and forests composed of mixed deciduous tree species.

Bioregion   geographical expanse which corresponds to the distribution of one or more groups of living beings, usually animals; for example, the Carolinian bioregion is characterized by the tulip tree, the raccoon, and so on.

Biosphere   that area where the atmosphere, lithosphere, and hydrosphere function together to form the context within which life exists.

Biotic   living; referring to Earth's living system of organisms.

Biotic (area)   see *Bioregion*.

Boreal forest   see *Tayga*.

Broad-leafed   with leaves other than linear in outline; as opposed to needle-leafed or grasslike (graminoid).

Brown forest soil (also called *gray-brown podzolic*)   acid soils with dark brown surface layers, rich in humus, grading through lighter colored soil layers to limy parent material; develops under deciduous forest.

Caliche   layer chiefly of calcium carbonate at or near ground surface; attributable to deposition by evaporation of groundwater; characteristic of arid and semiarid regions.

Catena   see *Toposequence*.

Channelization   the modification of river channels for the purpose of flood control, land drainage, navigation, and the reduction or prevention of erosion.

Chernozem   fertile, black or dark brown soil under prairie or grassland with lime layer at some depth between 0.6 and 1.5 m (2 and 5 ft).

Chestnut-brown   short-grass soil in subhumid to semiarid climate with dark brown layer at top, which is thinner and browner than in chernozem soils, that grades downward to a layer of lime accumulation.

Cistern   a large receptacle for storing water; especially, a tank in which rainwater is collected for use.

Clear-cutting   removal of virtually all the trees, large or small, in a stand in one cutting operation.

Climate   generalized statement of the prevailing weather conditions at a given place, based on statistics of a long period of record.

Climatic climax vegetation   relatively stable vegetation that is in equilibrium with the climate and soils of the site.

Climatic regime   seasonality of temperature and moisture.

Climax   relatively stable state of the vegetation.

Commercial forest land   an area that is capable of growing trees an average rate of 20 ft³/acre (4.4 m³/hectare) per year, and not reserved for other purposes (e.g., park).

Compensation factor   environmental conditions (e.g., high groundwater table) that allow the same species to be present in two different regions, but for different reasons.

Coniferous   trees bearing cones and commonly having needle-shaped leaves usually retained during the year; adapted to moisture deficiency due to frozen ground or soils that are not moisture-retentive.

Crustal plate   segment of the Earth's crust (brittle outermost rock layer) moving as a unit, in contact with adjacent plates along plate boundaries.

Cultivar   strain, variety, or race of plant which originated and is maintained under cultivation.

Cumulative effect   effect on the environment that results from the incremental impact of proposed action when added to other past, present, and reasonably foreseeable future actions.

Deciduous   woody plants, or pertaining to woody plants, that seasonally lose all of their leaves and become temporarily bare-stemmed.

Desert   supporting vegetation of plants so widely spaced, or sparse, that enough of the substratum shows through to give the dominant tone to the landscape.

Desert soil   shallow, gray soils containing little humus and excessive amounts of calcium carbonate at depths less than 30 cm.

Desertification   degradation of the plant cover and soil as a result of overuse, especially during periods of drought.

Dry steppe   with 6–7 arid months in each year.

Earth flow   a type of mass movement in which incoherent slope material becomes saturated with water and flows at moderate to very rapid speeds.

Ecogeographic   of or referring to the geographic dimensions of ecology.

Ecological design   design that minimizes environmentally destructive impacts by integrating itself with living processes.

Ecological land classification   an integrated approach in which land is divided into ecosystem units of various scales, or sizes.

Ecological landtype (see also *site*)   the lowest level of ecosystem identification and classification.

Ecoregion (also called *ecosystem region*)   a large area of similar climate where similar ecosystems occur on similar sites (those having the same landform, slope, parent material, and drainage characteristics); for example, beach ridges throughout the Subarctic ecoregion usually support a dense growth of black spruce or jack pine.

Ecosystem   an area of any size with an association of physical and biological components so organized that a change in any one component will bring about a change in the other components and in the operation of the whole system.

Ecosystem geography   the study of how and why ecosystems are distributed.

Ecosystem management   use of an ecological approach that blends social, physical, economic, and biological needs and values to ensure productive, healthy ecosystems.

Ecotone   transition zone between two communities.

Ecotype   species with wide geographic range that develop locally adapted populations having different limits of tolerance to environmental factors.

Edaphic   resulting from the character of the soil and surface.

Elevation   height of a point above sea level.

Elevational zonation (also called altitudinal *or* vertical zonation*)*   arrangement of climatic zones and associated soil and vegetation at different elevations on mountainsides.

Esker   narrow, often sinuous embankment of coarse gravel and boulders deposited in the bed of a meltwater stream enclosed in a tunnel within stagnant ice of an ice sheet.

Estuarine   associated with an estuary (i.e., a deep water tidal habitat and its adjacent tidal wetlands, which are usually semienclosed by land but have open, partly obstructed, or sporadic access to the open ocean, and in which ocean water is at least occasionally diluted from freshwater runoff from the land).

Evapotranspiration   the total water loss from land by the combined processes of evaporation and transpiration.

Exotic river   stream that flows across a region of dry climate and derives its discharge from adjacent uplands where a water surplus exists.

Exposure   the direction in which a slope faces. This has effects on the climate of the slope, in terms of total insolation received, exposure to rain-bearing winds, amount and duration of frost and snow cover, and so on.

Floodplain   that part of a valley floor over which a river spreads during seasonal or short-term floods.

Forb   broad-leaved herb, as distinguished from the grasses.

Forest   open or closed vegetation with the principal layer consisting of trees averaging more the 5 m (16 ft) in height.

Forest–steppe   intermingling of steppe and groves or strips of trees.

Forest–tundra   intermingling of tundra and groves or strips of trees.

Formation   a world vegetation type dominated throughout by plants of the same life-form.

Formative process   a set of actions and changes that occur in the landscape as a result of geomorphic, climatic, biotic, and cultural activities.

Fuel break   a wide strip cleared through forest or brush land to inhibit the spread of fire.

Geomorphic   of or pertaining to the form of the Earth's surface.

Germplasm   the substance of the germ cells by which hereditary characteristics are believed to be transmitted.

Global warming   the theory that the Earth's atmosphere is gradually warming due to the buildup of certain gasses, including carbon dioxide and methane, which are released by human activities. The increased levels of these gases cause added heat energy from the Earth to be absorbed by the atmosphere instead of being lost to space.

Gray-brown podzolic soil   acid soil under broad-leaf deciduous forest; has thin, organic layer over grayish brown, leached layer; layer of deposition is darker brown.

Great soil group   one of several families of soils having common features mostly attributable to the climatic and vegetation regimes.

Green design   a sustainable approach to design that incorporates such environmental issues as follows: efficient and appropriate use of land, energy, water, and other resources; protection of significant habitats, endangered species, archaeological treasures, and cultural resources; and integration of work, habitat, and agriculture. It supports human and natural communities while remaining economically viable.

Greenhouse effect   accumulation of heat in the lower atmosphere through the absorption of long-wave radiation from Earth's surface.

Groundwater table   the upper surface of a zone of saturation except where that surface is formed by an impermeable body.

Habitat   particular kind of environment in which a species or community lives.

Histosol   soil order consisting of soils that are organic.

Home range   area over which an animal ranges throughout the year.

Homestead tree   deciduous trees planted on southwest corners of buildings on the semiarid High Plains that provide microclimate comfort for the building interior.

Hydrologic culture   a civilization based on the control and distribution of water for irrigation, navigation, and human consumption.

Hydrologic cycle   the unending transfer of water from the oceans to the land (via the atmosphere), and vice versa (via rivers).

Igneous rock   a type of rock formed by the solidification of magma, either within the Earth's crust (intrusive rock) or at the surface (extrusive or volcanic rock).

Inceptisol   soil order consisting of soils with weakly differentiated horizons showing alteration of parent materials.

Indigenous (species)   a species which is native to a particular region; endemic.

Intermittent stream   streams which flow only part of the time, as after a rainstorm, during wet weather, or during only part of the year.

Invertebrate   any animal without backbone, or spinal column; the classification includes all animals except fishes, amphibians, reptiles, birds, and mammals.

Kame   hill that originated as mass of sand and gravel deposited against glacial ice by glacial meltwater.

Karstification   related to the formation of karst (i.e., landscape or topography dominated by surface features of limestone solution and underlain by a limestone cavern system).

Land   pertaining to the terrestrial part of the earth, as distinguished from sea and air.

Land capability   level of use an area can tolerate without sustaining permanent damage.

Land evaluation   the assessment of the suitability of land for use in agriculture, forestry, engineering, hydrology, regional planning, recreation, and so on.

Landform   see *Physiography*.

Landform relief   the difference in elevation between the ridge and adjoining valley.

Landscape   see *Landscape mosaic*.

Landscape ecosystem   see *Landscape mosaic*.

Landscape mosaic   as defined for use in this book: a geographic group of the smallest, or local, ecosystems (sites).

Landslide   a type of mass movement in which the material displaced retains its coherence as a single body as it moves over a clearly defined plane of sliding.

Latisol   reddish, infertile tropical soil in which silica has been leached out, leaving a kaolinitic clay with a high content of iron and aluminum hydroxides.

Legume   any of a large group of plants of the pea family, characterized by true pods enclosing seeds; because of their ability to store up nitrates, legumes are often plowed under to fertilize the soil.

Lichen   combinations of algae and fungi living together symbiotically; typically form tough, leathery coatings or crusts attached to rocks and tree trunks.

Lithology   the physical character of a rock.

Lithosequence   the spatial pattern of ecosystems resulting from the change in the character of the underlying rocks.

Macroclimate   climate that lies just beyond the local modifying irregularities of landform and vegetation.

Magma   mobile, high-temperature molten rock, usually of silicate mineral composition and with dissolved gases.

Mass movement   downslope, unit movement of a portion of the land's surface (i.e., a single landslide).

Meadow   closed herbaceous vegetation, commonly in stands of rather limited extent, or at least not usually applied to extensive grasslands.

Microclimate   climate at or near the ground surface, such as within the vegetation and soil layer.

Mixed forest   forest with both needle-leafed and broad-leafed trees.

Model   a simplified verbal, graphic, or mathematical description used to help understand a complex object.

Mollisol   soil order consisting of soils with a thick, dark-colored, surface-soil horizon, containing substantial amounts of organic matter (humus) and high-base status.

Monoculture   the raising of only one crop or product without using the land for other purposes.

Moraine   accumulation of rock debris deposited by a glacier.

Neotropical migratory bird   a bird that breeds, at least to some extent, in North America and spends the nonbreeding season in Mexico, Central America, the Caribbean, and/or South America.

Nival   related to the tops of high mountains with perpetual snowpack and ice.

Old-growth forest   forest that has not been cut or disturbed by humans for hundreds of years.

Oligotrophic   clear-water lake, containing little plankton, often deep and cold and with little thermal stratification, harboring rather poor flora and fauna.

Open woodland (also called *steppe forest* and *woodland–savanna*) open forest with lower layers also open, having the trees or tufts of vegetation discrete but averaging less than their diameter apart.

Oxisol   soil order consisting of soils that are mixtures principally of kaolin, hydrates oxides, and quartz.

Passive solar   systems that collect, move, and store heat using natural heat-transfer mechanisms such as conduction and air-convection currents.

Pattern-based design   based on an understanding the patterns of a region in terms of process and then applying these patterns to select suitable land-use locations.

Pediment   gently sloping, rock-floored land surface found at the base of the mountain mass or cliff in an arid region.

Perennial stream   stream that flows throughout the year and from source to mouth.

Physiognomy (of vegetation)   the outward, superficial appearance of vegetation, without necessary reference to structure or function, even less composition; for example, forest of Douglas fir, of Sitka spruce, of white spruce, of red fir, all have a similar physiognomy.

Physiography   landform (including surface geometry and underlying geologic material).

Physiographic region   area of similar geologic structure and topographic relief that has had a unified geomorphic history; for example, the Great Plains of Fenneman (1928).

Plant formation   one or more plant communities exhibiting a definite structure and physiognomy; a structural or physiognomic unit of vegetation; for example, a deciduous broad-leaf forest.

Pleistocene   geological epoch from about 2 million to 10,000 years ago, characterized by recurring glaciers; the Ice Age.

Podzol   soil order consisting of acid soil in which surface soil is strongly leached of bases and clays.

Potential natural vegetation   vegetation that would exist if nature were allowed to take its course without human interference.

Prairie   consisting of tall grasses, mostly exceeding 1 m (3.28 ft) in height, comprising the dominant herbs, with subdominant forbs (broad-leafed herbs).

Prescribed burn   planned use of fire in wild-land management in the United States.

Rainforest   a dense forest, comprising tall trees, growing in areas of very high rainfall.

Red-yellow podzol (also called *yellow forest soil*)   acid soil under broad-leaf deciduous or needleleaf evergreen forest developed in areas of humid subtropical climate.

Region   see *Ecoregion*.

Regional ecology   the ecological relationships that prevail over a region.

Regionalism   Art movement in which painters take their subjects and themes from their native surroundings.

Resiliency   the ability of an ecosystem to maintain the desired condition of diversity, integrity, and ecological processes following disturbance.

Riparian   related to or living on the bank of a river or lake; for example, cottonwood forest.

Roadology   the science or study of roads or journeys and, by extension, the study of how they are used, where they lead, and the landscapes of small towns, tourist courts, diners, and roadhouses.

Savanna   closed grass or other predominantly herbaceous vegetation with scattered or widely spaced woody plants usually including some low trees.

Scale   level of spatial resolution perceived or considered.

Sclerophyll or *sclerophyllous*   refers to plants with predominantly hard stiff leaves that are usually evergreen.

Selva   the tropical rainforest.

Semiarid   dry, with a shortage of moisture for much of the year, but not so dry as an arid area.

Semidesert (also called *half-desert*)   area of xerophytic shrubby vegetation with a poorly developed herbaceous lower layer (e.g., sagebrush).

Semievergreen forest (also called a *monsoon forest*)   where many, although not all, of the trees lose their leaves; adaptation to a dry season in the tropics.

Sense of place   the collection of meanings, feelings, beliefs, symbols, values, and feelings that individuals or groups associate with a particular locality.

Shrub   a woody plant less than 5 m high.

Sierozem soil   see *Desert soil*.

Silviculture   generally, the science and art of cultivating and managing forest crops based on a knowledge of silvics.

Site   the smallest, or local, ecosystem

Slope stability   ability of a slope to resist failure by landsliding.

Soil orders   those 11 soil classes forming the highest category in the classification of soils.

Soil slip (also call *debris slide*)   a form of mass movement involving the downslope movement of weathered material above the bedrock.

Spatial   having to do with space or area; place-to-place distribution.

Species   a group of organisms of the same kind which reproduce among themselves but are usually reproductively isolated from other groups of organisms.

Spectral signature   characteristic distribution of wavelengths reflected by a substance; can be used to distinguish different types of vegetation, soils, and land use.

Spodosol   soil order consisting of soils that have accumulations of amorphous materials in subsurface horizons; soils of the boreal forest.

Steppe (also called *short-grass prairie*)   open herbaceous vegetation, less than 1 m high, with the tufts or plants discrete yet sufficiently close together to dominate the landscape.

Succession   the replacement of one community of plants and animals by another.

Sustainable design   the process of prescribing compatible land uses and buildings based on the limits of a place, locally as well as regionally.

Sustainability   ability of an ecosystem to maintain ecological processes and functions, biological diversity, and productivity over time.

Taxonomy   the grouping of objects into classes based on similarity of characteristics.

Tayga (also spelled *taiga*)   a swampy, parklike savanna with needle-leaved (usually evergreen) low trees or shrubs; the northern circumpolar boreal forest.

Tectonic activity   process of bending (folding) and breaking (faulting) of crustal mountains, concentrated on or near plate boundaries.

Temperate   climates of mid-latitudes (from 30° to 60° latitude) with both a summer and winter.

Tidal barrage   man-made barrier in a estuary that allows the incoming tide to flow through the barrier; the outgoing tide is held behind the barrier forming a lake that is released through a power-generating station.

Topoclimate   climate of very small space; influenced by topography.

Topography   the description of the surface features of a place; a map which shows these features is know as a topographical map.

Toposequence   a change of a community with topography.

Tundra   slow-growing, low-formation, mainly closed vegetation of dwarf shrubs, graminoids, and cryptograms, beyond the subpolar or alpine tree line.

Tundra soil   cold, poorly drained, thin layers of sandy clay and raw humus; without distinctive soil profiles.

Ultisol   soil order consisting of soils with horizons of clay accumulation and low base supply.

Vernacular architecture   suggesting something countrified, home-made, traditional; as used in connection with architecture, it indicates the traditional rural or small-town dwelling, the dwelling of the farmer or craftsman or wage earner.

Watershed   area drained by a river or stream and its tributaries.

Weed   from the human perspective, a plant out of place.

Wetland   a biological community in an area of wet ground; areas of marsh, peatlands or water whether permanent or temporary, with water which is static or flowing, fresh or brackish.

Woodland   cover of trees whose crowns do not mesh, with the result that branches extend to the ground.

Xeriscape   landscaping design for conserving water that uses drought-resistant or drought-tolerant plants.

zone   all areas in which the zonal soils have the potential of supporting the same climatic climax plant association.

Notes

[1]Terms in bold are defined in the Glossary.

[2]First coined by the Canadian forest researcher Orie Loucks in 1962. See his "A forest classification for the Maritime Provinces, In *Proceedings of the Nova Scotian Institute of Science* 25 (Pt. 2): 85–167 (1962) with separate map at 1 in. equals 19 miles.

[3]The Barbed Wire Museum in La Crosse, Kansas displays over 500 varieties of barbed wire.

[4]Soil Taxonomy soil orders (USDA Soil Survey Staff 1975); described in the Glossary.

[5]Zones of latitude may be described as follows: from the equator to 30° are *low latitudes*; from 30° to 60° are the *middle latitudes*; from 60° to the poles are the *high latitudes*.

[6]Other methods of mapping zones at the global scale are those of Thornthwaite (1933), Holdridge (1947), and Walter and Box (1976). All methods appear to work better in some areas than in others and to have gained their own adherents. I chose the Köppen system as the basis for ecoregion delineation because it has become the international standard for geographical purposes.

[7]As quoted in Wells (1994, p. 5).

[8]Nancy Selover, Department of Geography, Arizona State University, telephone conversation, 27 August 2001.

[9]As quoted in Soule and Piper (1992).

[10]Gordon Warrington, soil scientist, U.S. Forest Service (now retired), personal communication.

[11]James Omernik, research geographer, U.S. Environmental Protection Agency, personal communication.

[12]Robert L. Thayer, professor of landscape architecture, University of California, Davis, undated class handout.

[13]Xeriscape is a registered trademark held by the National Xeriscape Council.

[14]Cook, K. 1997. Twice the beauty: Make your garden a haven for animals and birds. *Gardening How-To* Jan./Feb. 1997: 56–57. *Gardening How-To* is the official publication of the National Home Gardening Club, Minnetonka, MN.

[15]Regarded by colleagues as one of two or three most influential geologists in the Survey, Dave Love is the subject of John McPhee's *Rising from the Plains* (New York: Farrar, Straus, and Giroux, 1986).

[16]For more information on this problem see Debano, L.F. 1981. *Water Repellent Soils: A State-of-the-Art.* General Technical Report PSW-46. Berkeley, CA: Pacific Southwest Forest and Range Experiment Station, USDA Forest Service.

[17]For more information about this problem, please refer to Knopf (1986).

[18]William Elliot, Engineering Technology Project, Rocky Mountain Research Station, USDA Forest Service, Moscow, Idaho, telephone conversation, 2 October 2001.

[19]TAC/Centre Directors Working Group. 1993. The Ecoregional Approach to Research in the CGIAR. TAC Secretariat, FAO, Rome. Gryseels G.; Kassam A. 1994. Characterization and Implementation of the CGIAR Ecoregional Concept. Paper prepared for the IFPRI [International Food Policy Research Institute] Ecoregional/2020 Vision Workshop held in Airlie Conference Centre, Virginia, November 7–9, 1994.

[20]Sharma, M.; Brown, L.; Qureshi, A.; Garcia, M. 1996. Ecoregional Mapping Helps Pinpoint Areas Prone to Malnutrition. IFPRI Report 18 (2). World Bank, Washington, DC.

[21]Singh, A.; Ramachandran, B.; Fosnight, G; et al. n.d. Biodiversity-rich Ecoregions in Africa Need Protection. Information for Decision Making Series. U.S. Geological Survey, EROS Data Center, Sioux Falls, SD.

[22]Barham, E. 1997. What's in a name: Eco-labelling in the global food system. Paper presented at the Joint Meeting of the Agriculture, Food, and Human Values Society and the Association for the Study of Food and Society, held in Madison, Wisconsin, June 5–8, 1997.

[23.]Starbucks. 2000. Commitment to origins: Starbucks involvement in coffee-origin countries. Seattle, WA: Starbucks.

[24]As far as I know, the priorities have not changed. However, I found a reference to an unpublished paper by A.F. McCalla of The World Bank, dated 1991, with the intriguing title "Ecoregional Basis for International Research Investment." Also, an abstract of a paper by Roger Sayre and Xiaojun Li of The Nature Conservancy titled "An Ecoregional Conservation Strategy for Latin America and the Caribbean," given at the 1997 ESRI (Environment Systems Research Institute) User Conference, is pertinent to the current situation. They state ". . . The World Bank and the United States Agency for International Development (USAID) have emphasized the delineation of ecoregions . . . , and an assignment of biodiversity importance values to these ecoregions. The World Bank and USAID are now using these ecoregion priorities assessments to help determine project placements and conservation allocations."

Selected Bibliography

Aberley, D. (ed.). 1994. *Futures by Design: The Ecological Practice of Ecological Planning.* Gabriola Island, BC: New Society Publishers.

Albert, D.A. 1995. *Regional Landscape Ecosystems of Michigan, Minnesota, and Wisconsin: A Working Map and Classification.* General Technical Report NC-178. St. Paul, MN: North Central Forest Experiment Station. With separate map, scale = 1 : 2,000,000.

Aldrich, J.W.; James, F.C. 1991. Ecogeographic variation in the American Robin. *The Auk* 108:230–249.

Alexander, C.; Ishikawa, S.; Silverstein, M., et al. 1977. *A Pattern Language.* New York: Oxford University Press.

Allen, T.F.H.; Hoekstra, T.W. 1992. *Toward a Unified Ecology.* New York: Columbia University Press.

Anon. 1993. Studying climate-controlled "ecoregions" to understand forests. *Forest Perspectives* 3(3): 4.

Atwood, W.W. 1940. *The Physiographic Provinces of North America.* Boston: Ginn. With separate map of "Landforms of the United States" by Erwin Raisz at 1 in. equals 75 miles.

Bailey, R.G. 1971. *Landslide Hazards Related to Land Use Planning in Teton National Forest, Northwest Wyoming.* Ogden, UT: USDA Forest Service, Intermountain Region. With separate map at one-half in. equal 1 mile.

Bailey, R.G. 1974. *Land-capability Classification of the Lake Tahoe Basin, California–Nevada: A Guide for Planning.* South Lake Tahoe, CA: USDA Forest Service in cooperation with the Tahoe Regional Planning Agency. With separate map at three-quarter in. equal 1 mile.

Bailey, R.G. 1976. *Ecoregions of the United States.* Ogden, UT: Intermountain Region, USDA Forest Service. Map, scale = 1 : 7,500,000.

Bailey, R.G. 1983. Delineation of ecosystem regions. *Environmental Management* 7:365–373.

Bailey, R.G. 1984. Testing an ecosystem regionalization. *Journal of Environmental Management* 19:239–248.

Bailey, R.G. 1985. The factor of scale in ecosystem mapping. *Environmental Management* 9:271–276.

Bailey, R.G. 1987. Suggested hierarchy of criteria for multi-scale ecosystem mapping. *Landscape and Urban Planning* 14:313–319.

Bailey, R.G. 1988a. *Ecogeographic Analysis: A Guide to the Ecological Division of Land for Resource Management.* Washington, DC: USDA Forest Service. Misc. Publ. 1465.

Bailey, R.G. 1988b. Problems with using overlay mapping for planning and their implications for geographic information systems. *Environmental Management* 12:11–17.

Bailey, R.G. 1989. Ecoregions of the Continents—Scale 1 : 30,000,000 [map of land-masses of the world, published as a] supplement in *Environmental Conservation* 16(4), and accompanying Explanatory Supplement to Ecoregions of the Continents (pp. 307–309).

Bailey, R.G. 1991. Design of ecological networks for monitoring global change. *Environmental Conservation* 18:173–175.

Bailey, R.G. 1995. *Description of the Ecoregions of the United States.* 2nd ed. rev. and expanded. Washington, DC: USDA Forest Service. Misc. Publ. No. 1391 (rev.). With separate map, scale = 1 : 7,500,000.

Bailey, R.G. 1996. *Ecosystem Geography.* New York: Springer-Verlag.

Bailey, R.G. 1998a. *Ecoregions: the Ecosystem Geography of the Oceans and Continents.* New York: Springer-Verlag.

Bailey, R.G. 1998b. *Ecoregions Map of North America: Explanatory Note.* Washington, DC: USDA Forest Service. Misc. Publ. 1548. With separate map, scale = 1 : 15,000,000. In cooperation with The Nature Conservancy and the U.S. Geological Survey.

Bailey, R.G.; Hogg, H.C. 1986. A world ecoregions map for resource reporting. *Environmental Conservation* 13:195–202.

Bailey, R.G.; Rice, R.M. 1969. Soil slippage: an indicator of slope instability on chaparral watersheds of southern California. *Professional Geographer* 21:172–177.

Bailey, R.G.; Avers, P.E.; King, T.; McNab, W.H. (compilers and eds.). 1994. *Ecoregions and Subregions of the United States.* Washington, DC: USDA Forest Service. Map, scale = 1 : 7,500,000; with supplementary table of map unit descriptions, compiled and edited by W.H. McNab and R.G. Bailey.

Bailey, R.W. 1941. Climate and settlement of the arid region. In G. Hambidge (ed.). *Climate and Man, 1941 Yearbook of Agriculture.* Washington, DC: U.S. Department of Agriculture, pp. 188–196.

Barnes, B.V. 1984. Forest ecosystem classification and mapping in Baden–Wurttenberg, Germany. In J.G. Bockheim (ed.). *Proceedings of the Symposum Forest Land Classification: Experience, Problems, Perspectives.* NCR-102 North Central Forest Soils Conference, Society of American Foresters, USDA Forest Service and USDA Soil Conservation Service, Madison, WI, pp. 49–65.

Barnes, B.V.; Pregitzer, K.S.; Spies, T.A.; Spooner, V.H. 1982. Ecological forest site classification. *Journal of Forestry* 80:493–498.

Barnes, B.V.; Zak, D.R.; Denton, S.R.; Spurr, S.H. 1998. *Forest Ecology,* 4th ed. New York: John Wiley & Sons.

Barnett, D.L. with Browning, W.D. [illustrations by J.L. Uncapher] 1995. *A Primer on Sustainable Building.* Snowmass, CO: Rocky Mountain Institute.

Bennett, C.F. 1975. *Man and Earth's Ecosystems.* New York: John Wiley & Sons.

Berghaus, H. 1845. *Physikalischer Atlas.* Gotha, Germany: [publisher unknown]. Vol. 1.

Bhat G.; Bergstrom, J.; Bowker, J.M.; Cordell, H.K. 1998. An ecoregional approach to the economic valuation of land and water based recreation in the United States. *Environmental Management* 22:69–77.

Borchert, J.F. 1950. The climate of the central North American grassland. *Annals Association of American Geographers.* 40:1–39.

Bormann, F.H; Balmori, D.; Geballe, G.T. 1993. *Redesigning the American Lawn: A Search for Environmental Harmony.* New Haven, CT: Yale University Press.

Bouma, J., et al. (eds.). 1995. *Eco-regional Approaches for Sustainable Land Use and Food Production.* Dordrecht: Kluwer Academic Publishers in cooperation with International Potato Center.

Bowers, J.E. 1988. *A Sense of Place: The Life and Work of Forrest Shreve.* Tucson, AZ: University of Arizona Press.

Bowman, I. 1916. *The Andes of Southern Peru: Geographical Reconnaissance Along the Seventy-third Meridian.* New York: American Geographical Society by Henry Holt and Co.

Brand, S. 1994. *How Buildings Learn: What Happens After They're Built.* New York: Viking Press.

Branson, F.A.; Shown, L.M. 1989. *Contrasts of Vegetation, Soils, Microclimates, and Geomorphic Processes Between North- and South-facing Slopes on Green Mountain Near Denver, Colorado.* Report 89-4094. Denver: U.S. Geological Survey Water Resources Investigations.

Brewer, I. 1999. *The Conceptual Development and Use of Ecoregion Classification.* Master's thesis. Corvallis, OR: Oregon State University.

Breymeyer, A.I. 1981. Monitoring of the functioning of ecosystems. *Environmental Monitoring and Assessment* 1:175–183.

Brown, J.F.; Loveland, T.R.; Merchant, J.W.; Reed, B.C.; Ohlen, D.O. 1993. Using mulitsource data in global land-cover characterization: Concepts, requirement, and methods. *Photogrammetric Engineering & Remote Sensing* 59:977–987.

Brown, J.H. 1995. *Macroecology.* Chicago: University of Chicago Press.

Bryson, B. 1990. *The Lost Continent: Travels in Small Town America.* New York: HarperCollins.

Bunnett. R.B. 1968. *Physical Geography in Diagrams.* New York: Frederick A. Praeger.

Burger, D. 1976. The concept of ecosystem region in forest site classification. In *Proceedings, International Union of Forest Research Organizations (IUFRO).* XVI World Congress, Division I: 20 June–2 July 1976; Oslo, Norway. Oslo: IUFRO; pp. 213–218.

Burger, D. 1986. Physiography as an integral part of forest ecosystems. In Y. Hanxi, W. Zhan, J.N.R. Jeffers, and P.A. Ward (eds.). *Proceedings, International Symposium on Temperate Forest Ecosystem Management and Envi-*

ronmental Protection. Antu, Jilin Province, China. Cumbria, United Kingdom: Institute of Terrestrial Ecology.

Busch, A. 1999. *Geography of Home: Writing on Where We Live*. New York: Princeton Architectural Press.

Byers, B. 1991. Ecoregions, state sovereignty and conflict. *Bulletin of Peace Proposals* 22(1):65–76.

Campbell, R.H. 1975. *Soil Slips, Debris Flows, and Rainstorms in the Santa Monica Mountains and Vicinity, Southern California*. U.S. Geological Survey Professional Paper 851.

Christopherson, R.W. 2000. *Geosystems: An Introduction to Physical Geography*. 4th ed. Upper Saddle River, NJ: Prentice-Hall.

Cleland, D.T.; Avers, P.E.; McNab, W.H.; Jensen, M.E.; Bailey, R.G.; King, T.; Russell, W.E.. 1997. National hierarchical framework of ecological units. In M.S. Boyce and A. Haney (eds.). *Ecosystem Management: Applications for Sustainable Forest and Wildlife Resources*. New Haven, CT: Yale University Press, pp. 181–200.

Clements, F. 1916. *Plant Succession: An Analysis of the Development of Vegetation*. Publication 242. Washington, DC: Carnegie Institution of Washington.

Clements, F.; Shelford, V.E. 1939. *Bioecology*. New York: John Wiley & Sons.

Commission for Environmental Cooperation. 1997. *Ecological Regions of North America: Towards a Common Perspective*. Quebec, Canada: Commission for Environmental Cooperation.

Common, R. 1966. Slope failure and morphogenetic regions. In G.H. Dury (ed.). *Essays in Geomorphology*. New York: American Elsevier, pp. 53–82.

Corbel, J. 1964. L'erosion terrestre etude quantitative (methods-technques-resultats). *Annales de Geographie* 73:385–412.

Corn, W.M. 1983. *Grant Wood: The Regionalist Vision*. New Haven, CT: Yale University Press.

Cowardin, L.M.; Carter, V.; Golet, F.C.; LaRoe, E.T. 1979. *Classification of Wetlands and Deep-Water Habitats of the United States*. Report No. FWS/OBS-79/31. Washington, DC: U.S. Fish and Wildlife Service.

Cox, J. 1991. *Landscaping with Nature: Using Nature's Design to Plan Your Yard*. Emmaus, PA: Rodale Press.

Crowley, J. 1967. Biogeography. *Canadian Geographer* 11:312–326.

Daniels, S. 1995. *The Wild Lawn Handbook: Alternatives to the Traditional Front Lawn*. New York: Macmillan.

Dansereau. P. 1957. *Biogeography—An Ecological Perspective*. New York: Ronald Press.

Dasmann, R.F. 1973. *A System for Defining and Classifying Natural Regions for Purposes of Conservation*. Morges, Switzerland: International Union for Conservation of Nature and Natural Resources (IUCN). IUCN Occasional Paper No. 8.

Daubenmire. R. 1968. *Plant Communities: A Text Book on Plant Synecology*. New York: Harper & Row.

Davis, M. 1998. *Ecology of Fear: Los Angeles and the Imagination of Disaster*. New York: Vintage Books.

Davis, N.D. 1994. Following the mapmaker. *American Forests*. May/June: 23–25, 58–59.

DeLano, P.; Johnson, C. 1993. *Kansas: Off the Beaten Path,* 2nd ed. Old Saybrook, CT: Globe Pequot Press.

Denniston, D. 1995. *High Priorities: Conserving Mountain Ecosystems and Cultures.* Worldwatch Paper 123. Washington, DC: Worldwatch Institute.

Dice, L.R. 1943. *The Biotic Provinces of North America.* Ann Arbor: University of Michigan Press.

Doe, W.W.; Shaw, R.B.; Bailey, R.G.; Jones, D.S.; Macia, T.E. 2000. U.S. Army training and testing lands: An ecoregional framework for assessment. In E.J. Palka and F.A. Galgano (eds.). *The Scope of Military Geography: Across the Spectrum from Peacetime to War.* New York: McGraw-Hill, pp. 373–392.

Dokuchaev, V.V. 1899. *On the theory of natural zones. Sochineniya Vol. 6.* Moscow: Academy of Science of the USSR, reprinted 1951.

Dranstad, W.E.; Olson, J.D.; Forman, R.T.T. 1996. *Landscape Ecology Principles in Landscape Architecture and Land-use Planning.* Washington, DC: Island Press.

Ecoregions Working Group. 1989. *Ecoclimatic Regions of Canada, First Approximation.* Ecological Land Classification Series No. 23. Ottawa: Environment Canada. With separate map, scale = 1 : 7,500,000.

Elder, J. 1994. The big picture: Sierra Club Critical Ecoregions Program. *Sierra* Mar./April: 52–57.

Fenneman, N.M. 1928. Physiographic divisions of the United States. *Annals Association of American Geographers* 18:261–353.

Fisk, P. 1983. *Bioregions and Biotechnologies: A New Planning Tool for Stable State Economic Development.* Austin, TX: Center for Maximum Potential Building Systems.

Foreign Agricultural Organization of the United Nations (FAO). 2000. *Global Ecological Zoning for the Forest Resources Assessment 2000.* Rome: Foreign Agricultural Organization of the United Nations, Forestry Department.

Forman, R.T.T. 1995. *Land Mosaic: The Ecology of Landscapes and Regions.* Cambridge: Cambridge University Press.

Foster, C.H.W. 1997. Bioregionalism revisited. *Renewable Resources Journal* 15(3):6–10.

Fournier, F. 1960. *Climat et Erosion.* Paris: Presses Universite France.

Francis, M.; Reimann, A.; Nascimbene, Y. [illustrator]. 1999. *The California Landscape Garden: Ecology, Culture and Design.* Berkeley: University of California Press.

Franklin, J.F. 1993. Preserving biodiversity: Species, ecosystems, or landscapes? *Ecological Applications* 3:202–205.

Gallant, A.L.; Binnian, E.F.; Omernik, J.M.; Shasby, M.B. 1995. *Ecoregions of Alaska.* U.S. Geological Survey Professional Paper 1567. Washington, DC: U.S. Government Printing Office. With separate map, scale = 1 : 5,000,000.

Gersmehl, P.; Napton, D.; Luther, J. 1982. The spatial transferability of resource interpretations. In T.B. Braun (ed.). *Proceedings, National In-place Resource Inventories Workshop.* Washington, DC: Society of American Foresters, pp. 402–405.

Glausiusz, J. 1997. The ecology of language. *Discover* Aug: 30.

Goudey, C.B.; Smith, D.W. (compilers and editors). 1994. *Ecological Units of California: Subsections.* San Francisco, CA: USDA Forest Service California

Region. Map, scale = 1 : 1,000,000. In cooperation with the Natural Resources Conservation Service.

Goudie, A.; Viles, H. 1997. *The Earth Transformed: An Introduction to Human Impacts on the Environment*. Oxford: Blackwell.

Günther, M. 1955. Untersuchingen über das Ertragsvermögen der Hauptholzarten in Bereich verschiedener des württembergischen Neckarlandes. *Mitt. Vereins f. Forstl. Standortsk. u. Forstpflz.* 4:5–31.

Hack, J.T.; Goodlet, J.C. 1960. *Geomorphology and Forest Ecology of a Mountain Region in the Central Appalachians*. U.S. Geological Survey Professional Paper 347. Washington, DC: U.S. Geological Survey.

Hammond, E.H. 1954. Small-scale continental landform maps. *Annals Association of American Geographers*. 44:33–42.

Hart, J.F. 1998. *The Rural Landscape*. Baltimore, MD: John Hopkins University Press.

Herbertson, A.J. 1905. The major natural regions: An essay in systematic geography. *Geography Journal* 25:300–312.

Hills, G.A. 1952. *The Classification and Evaluation of Site for Forestry*. Research Report 24. Toronto: Ontario Department of Lands and Forest.

Hills, G.A. 1960. Comparison of forest ecosystems (vegetation and soil) in different climatic zones. *Silva Fennica* 105:33–39.

Hills, G.A. 1961. *The Ecological Basis for Land-use Planning*. Research Report No. 46. Toronto: Ontario Department of Lands and Forests

Holdridge, L.R. 1947. Determination of world plant formations from simple climatic data. *Science* 105:367–368.

Host, G.E.; Pregitzer, K.S.; Ramm, C.W.; Hart, J.B.; Cleland, D.T. 1987. Landform-mediated differences in successional pathways among upland forest ecosystems in northwestern Lower Michigan. *Forest Science* 33:445–457.

Hough, M. 1990. *Out of Place: Restoring Identity to the Regional Landscape*. New Haven: Yale University Press.

Hough, M. 1994. Design with city nature. In R.H. Platt, R.A. Rountree, and P.C. Muich (eds.). *The Ecological City: Perserving and Restoring Urban Biodiversity*. Amherst: University of Massachusetts Press, pp. 40–48.

Huang, S.; Price, D; Titus, S.J. 2000. Development of ecoregion-based height-diameter models for white spruce in boreal forests. *Forest Ecology and Management* 129:125–141.

Hudson, J.C. 1985. *Plains Country Towns*. Minneapolis: University of Minnesota Press.

Huggett, R.J. 1995. *Geoecology: An Evolutionary Approach*. London: Routledge.

Hughes, R.M. 1997. Use of ecoregions in biological monitoring. In S.L. Loeb and A. Spacie (ed.). *Biological Monitoring of Aquatic Systems*. Boca Raton, FL: Lewis Publishers, pp. 125–151.

Hunsaker, C.T., et al. 1990. Assessing ecological risk on a regional scale. *Environmental Management* 14:325–332.

Hunt, C.B. 1974. *Natural Regions of the United States and Canada*. San Francisco, CA: W.H. Freeman.

Illies, J. 1974. *Introduction to Zoogeography* [transl. from German by W.D. Williams]. London: Macmillan.

Isachenko, A.G. 1973. *Principles of Landscape Science and Physical–*

Geographical Regionalization [transl. from Russian by R.J. Zatorski] J.S. Massey (ed.). Carlton, Victoria, Australia: Melbourne University Press.

Jackson, J.B. 1984. *Discovering the Vernacular Landscape.* New Haven, CT: Yale University Press.

Jackson, J.B. 1994. *A Sense of Place, a Sense of Time.* New Haven, CT: Yale University Press.

Jackson, W. 1994. *Becoming Native to This Place.* Lexington: University of Kentucky Press.

James, P.E. 1952. Toward a further understanding of the regional concept. *Annals Association of American Geographers* 42:195–222.

James, P.E. 1959. *A Geography of Man,* 2d ed. Boston: Ginn.

Johnson, N.C. [and others] (eds.). 1999. *Ecological Stewardship: A Common Reference for Ecosystem Management*: 3 volume set. Oxford: Elsevier Science.

Keys, J.E.; Carpenter, C.A.; Hooks, S.L.; Koeneg, F.G.; McNab, W.H.; Russell, W.E.; Smith, M.L. 1995. *Ecological Units of the Eastern United States, First Approximation.* Technical Publication R8-TP21. Atlanta, GA: USDA Forest Service. Map, scale = 1 : 3,500,000.

Kiser, M.; Kennedy, J. 1997. Ecoregion analysis—Looking at bat houses in their environment. *The Bat House Researcher* 5(1):6–7.

Klijn, F.; Udo de Haes, H.A. 1994. A hierarchical approach to ecosystems and its implications for ecological land classification. *Landscape Ecology* 9:89–104.

Klijn, F.; DeWaal, R.W.; Oude Voshaar, J.H. 1995. Ecoregions and ecodistricts: Ecological regionalization for the Netherlands Environmental policy. *Environmental Management* 19:797–813.

Klopatek, J.M. et al. 1979. *The Need for Regional Ecology.* Environmental Sciences Division Publication No. 1318. Oak Ridge, TN: Oak Ridge National Laboratory.

Knight, R.L.; Landres, P.B. (eds.). 1998. *Stewardship Across Boundaries.* Washington, DC: Island Press.

Knopf, F.L. 1986. Changing landscapes and the composition of the eastern Colorado avifauna. *Wildlife Society Bulletin* 14(2):132–142.

Köppen, W. 1931. *Grundriss der Klimakunde.* Berlin: Walter de Gruyter.

Küchler, A.W. 1964. *Potential Natural Vegetation of the Conterminous United States.* Special Publication 36. New York: American Geographical Society. With separate map, scale = 1 : 3,168,000.

Kunstler, J.H. 1993. *The Geography of Nowhere: The Rise and Decline of America's Man-made Landscape.* New York: Simon & Schuster.

Kunstler, J.H. 1996. *Home from Nowhere: Remaking our Everday World for the 21st Century.* New York: Simon & Schuster.

Lawrence, R.G. 2000. The godparents of green. *Natural Home* May/June: 44–51.

Leser, H.; Nagel, P. 1998. Landscape diversity—A holistic approach. In W. Barthloff and M. Wininger (eds.). *Biodiversity. A Challenge for Development, Research and Policy.* Berlin: Springer-Verlag, pp. 129–143.

Logan, T.L. 1985. An ecoregion-continuum approach to global vegetation biomass estimation. In *Proceedings, Tenth W.T. Pecora Memorial Remote Sensing Symposium.* Falls Church, VA: American Society of Photogrammetry and Remote Sensing, pp. 483–493.

Loomis, J.; Echohawk, J.C. 1999. Using GIS to identify under-represented ecosystems in the National Wilderness Preservation System in the USA. *Environmental Conservation* 26:53–58.

Luccarelli, M. 1995. *Lewis Mumford and the Ecological Region: The Politics of Planning.* New York: Guilford Press.

Lyle, J.T. 1994. *Regenerative Design for Sustainable Development.* New York: John Wiley & Sons.

Lyle, J.T. 1999. *Design for Human Ecosystems: Landscape, Land Use, and Natural Resources.* Washington, DC: Island Press. (Originally published in 1985 by Van Nostrand Reinhold.)

Lynch, K. 1976. *Managing the Sense of a Region.* Cambridge, MA: MIT Press.

MacArthur, R.H. 1972. *Geographical Ecology: Patterns in the Distribution of Species.* New York: Harper & Row.

Mackinnon, A; Meidinger, D.; Klinka, K. 1992. Use of biogeoclimatic ecosystem classification system in British Columbia. *Forestry Chronicle* 68:100–120.

Major, J. 1951. A functional, factorial approach to plant ecology. *Ecology* 32: 392–412.

Marr, J.W. 1961. *Ecosystems of the East Slope of the Front Range in Colorado.* Boulder: University of Colorado Press.

Marsh, W.M. 1998. *Landscape Planning: Environmental Applications,* 3rd ed. New York: John Wiley & Sons.

Maser, C.; Sedell, J.R. 1994. *From the Forest to the Sea: The Ecology of Wood in Streams, Rivers, Estuaries, and Oceans.* Delray Beach, FL: St. Lucia Press.

Mason, R.J.; Mattson, M.T. 1990. *Atlas of United States Environmental Issues.* New York: Macmillan.

Mather, J.R.; Sdasyuk, G.V. (eds.). 1991. *Global Change: Geographical Approaches.* Tucson: University of Arizona Press.

Matthews, A. 1992. *Where the Buffalo Roam.* New York: Grove Press.

McGinnis, M.V. (ed.). 1999. *Bioregionalism.* New York: Routledge.

McHarg, I. 1969. *Design with Nature.* Garden City, NY: Natural History Press.

McHarg, I. 1997. Natural factors in planning. *Journal of Soil and Water Conservation* 52(1):13–17.

McHarg, I.L.; Steiner, F.R. (eds.). 1998. *To Heal the Earth: Selected Writings of Ian L. McHarg.* Washington, DC: Island Press.

McHarg, I.; Sutton, J. 1975. Ecological planning for the Texas Coastal Plain. *Landscape Architecture* 65(1):78–89.

McNab, W.H. 1990. Predicting forest type in Bent Creek Experimental Forest from topographic variables. In S.J. Coleman and D.G. Neary (eds.). *Proceeding of the Sixth Biennial Southern Silvicultural Research Conference.* General Technical Report SE-70. Asheville, NC: USDA Forest Service, Southeastern Forest Experiment Station.

McNab, W.H.; Avers, P.E. 1994. *Ecological Subregions of the United States: Section Descriptions.* Administrative Publ. WO-WSA-5. Washington, DC: USDA Forest Service.

McMahon, G.; Gregonis, S.M.; Waltman, S.W.; et al. 2001. Developing a spatial framework of common ecological regions for the conterminous United States. *Environmental Management* 28:293–316.

Merriam, C.H. 1898. *Life Zones and Crop Zones of the United States.* Bulletin

Division Biological Survey 10. Washington, DC: U.S. Department of Agriculture, pp. 1–79.

Mittermeier, R.A.; Myers, N.; Mittermeier, C.G. 2000. *Hotspots: Earth's Biologically Richest and Most Endangered Terrestrial Ecoregions.* Chicago: University of Chicago Press.

Milanova, E.V.; Kushlin, A.V. (eds.). 1993. *World Map of Present-day Landscapes: An Explanatory Note.* Moscow: Moscow State University. With separate map, scale = 1 : 15,000,000.

Miller, K.R. 1996. *Balancing the Scales: Guidelines for Increasing Biodiversity's Chances Through Bioregional Management.* Washington, DC: World Resources Institute.

Minnich, R.A. 1988. *The Biogeography of Fire in the San Bernardino Mountains of California: A Historical Study.* University of California Publications in Geography Vol. 28. Berkeley: University of California Press.

Mollison, B., with Renay Mia Slay.1994. *Introduction to Permaculture,* 2nd ed. Tyalgum, Australia: Tagari Publications.

Moon, William Least Heat [William Trogdon] 1982. *Blue Highways: A Journey into America.* Boston: Little, Brown.

National Park Service, U.S. Department of Interior. 1993. *Guiding Principles of Sustainable Design.* Denver, CO: National Park Service, Denver Service Center.

National Resources Committee. 1935. *Regional Factors in National Planning and Development.* Washington, DC: U.S. Government Printing Office.

Nature Conservancy, The. 1997. *Designing a Geography of Hope: Guidelines for Ecoregion-based Conservation in The Nature Conservancy.* Arlington, VA: The Nature Conservancy.

Nesser, J.A.; Ford, G.L.; Maynard, C.L.; Page-Dumroese, D.S. 1997. *Ecological Units of the Northern Region: Subsection.* General Technical Report INT-GTR-369. Ogden, UT: Intermountain Research Station.

Newell, W.L. 1978. *Understanding Natural Systems—A Perspective for Land-use Planning in Appalachian Kentucky.* U.S. Geological Survey Bulletin 1438. Washington, DC: U.S. Geological Survey.

Noss, R. 1983. A regional approach to maintain diversity. *BioScience* 33:700–706.

Noss, R. 1999/2000. A reserve design for the Kalamath–Siskiyou ecoregion. *Wild Earth* 9(4):71–76.

O'Brien, R.A. 1996. *Forest Resources of Northern Utah Ecoregions.* Resource Bulletin INT-RB-87. Ogden, UT: USDA Forest Service Intermountain Research Station.

Odum, E.P. 1971. *Fundamentals of Ecology,* 3rd ed. Philadelphia: W.B. Saunders.

Odum, E.P. 1977. The emergence of ecology as a new integrative discipline. *Science* 195:1289–1293.

Odum, E.P. 1992. Great ideas in ecology for the 1990s. *Bioscience* 42:524–545.

Olgyay, V. 1963. *Design with Climate: Bioclimatic Approach to Architectural Regionalism.* Princeton, NJ: Princeton University Press.

Olsen, D.M.; Dinerstein, E. 1998. The Global 200: A representative approach to conserving the Earth's most biologically valuable ecoregions. *Conservation Biology* 12:502–515.

Omernik, J.M. 1987. Map supplement: Ecoregions of the conterminous United States. *Annals Association of American Geographers* 77:118–125. Scale = 1 : 7,500,000.

Omernik, J.M.; Bailey, R.G. 1997. Distinguishing between watersheds and ecoregions. *Journal American Water Resources Association* 33:1–15.

Opie, J. 1993. *Ogallala: Water for a Dry Land.* Lincoln: University of Nebraska Press.

Opie, J. 1998. *Nature's Nation: An Environmental History of the United States.* Fort Worth, TX: Harcourt Brace College Publishers.

Orme, A.T.; Bailey, R.G. 1971. Vegetation and channel geometry in Monroe Canyon, southern California. *Yearbook of the Association Pacific Coast Geographers* 33:65–82.

Orr, D.L. 1992. *Ecological Literacy: Education and the Transition to a Postmodern World.* Albany: State University of New York Press.

Passarge, S. 1929. *Die Landschaftsgurtel der Erde, Natur und Kultur.* Breslau: Ferdinand Hirt.

Peet, R.K. 1988. Forests of the Rocky Mountains. In M.G. Barbour and W.D. Billings (eds.). *North American Terrestrial Vegetation.* Cambridge: Cambridge University Press, pp. 63–102.

Pflieger, W.L. 1971. *A Distributional Study of Missouri Fishes.* University of Kansas: Museum of Natural History Publication 20:225–570.

Phillips, J. 1995. *Natural by Design: Beauty and Balance in Southwest Gardens.* Santa Fe: Museum of New Mexico Press.

Pollan, M. 1991. *Second Nature: A Gardener's Education.* New York: Atlantic Monthly Press.

Popper, D.; Popper, F. 1987. The Great Plains: From dust to dust. *Planning* 53(12):12–18.

Pringle, C.M. 1997. Exploring how disturbance is transmitted upstream: Going against the flow. *Journal of the North American Benthological Society* 16:425–438.

Pyne, S.J. 1991. *Burning Bush: A Fire History of Australia.* New York: Holt.

Quigley, T.M., et al. 1998. Using an ecoregion assessment for integrated policy analysis. *Journal of Forestry* 96(10):33–38.

Radbruch-Hall, D.H.; Colton, R.B.; Davies, W.E.; et al. 1982. *Landslide Overview Map of the Conterminous United States.* U.S. Geological Survey Professional Paper 1183. Washington, DC: U.S. Geological Survey. With separate map, scale = 1 : 7,500,000.

Reiniger, C. 1997. Bioregional planning and ecosystem protection. In G.F. Thompson and F. Steiner (eds.). *Ecological Design and Planning.* New York: John Wiley & Sons, pp. 185–200.

Reisner, M. 1986. *Cadillac Desert: The American West and its Disappearing Water.* New York: Viking Penguin.

Retzer, J.L. 1965, Significance of stream systems and topography in managing mountain lands. In C.T. Youngberg (ed.). *Forest-soil Relationships in North America.* Corvallis: Oregon State University Press, pp. 399–411.

Ricketts, T.A., et al. 1999. *Terrestrial Ecoregions of North America: A Conservation Assessment.* Washington, DC: Island Press.

Riebsame, W. 1991. Sustainability of the Great Plains in an uncertain climate. *Great Plains Research* 1:133–151.

Robertson, J.K.; Wilson, J.W. 1985. *Design of the National Trends Network for Monitoring the Chemistry of Atmospheric Precipitation.* U.S. Geological Circular 964. Washington, DC: U.S. Geological Survey.

Roca, R.; Adkins, L.; Wurschy, M.C.; Skerl, K.L. 1996. *Wings from Afar: An Ecoregional Approach to Conservation of Neotropical Migratory Birds in South America.* Arlington, VA: The Nature Conservance.

Rosgen, D. [illustration by Silvey, H.L.]. 1996. *Applied River Morphology.* Pagosa Springs, CO: Wildland Hydrology.

Rowe, J.S. 1979. Revised working paper on methodology/philosophy of ecological land classification in Canada. In C.D.A. Rubec (ed.). *Applications of Ecological (Biophysical) Land Classification in Canada.* Ecological Land Classification Series No. 7. Ottawa: Environment Canada. pp. 23–30.

Rowe, J.S. 1980. The common denominator in land classification in Canada: An ecological approach to mapping. *Forestry Chronicle* 56:19–20.

Rowe, J.S.; Sheard, J.W. 1981. Ecological land classification: A survey approach. *Environmental Management* 5:451–464.

Rudis, V.A. 1998. Regional forest resource assessment in an ecological framework: The Southern United States. *Natural Areas Journal* 18:319–332.

Rudis, V.A. 1999. *Ecological Subregion Codes by County, Coterminous United States.* General Technical Report SRS-36. Asheville, NC: USDA Forest Service, Southern Research Station.

Sacks, J.D.; Mellinger, A.D.; Gallup, J.L. 2001. The geography of poverty and wealth. *Scientific American.* 284(3):70–75.

Sanders, R.A.; Rowntree, R.A. 1983. *Classification of American Metropolitan Areas by Ecoregion and Potential Natural Vegetation.* Research Paper NE-516. Broomall, PA: USDA Forest Service, Northeastern Forest Experiment Station.

Schmithusen, J. 1976. *Atlas zur Biogeographie.* Mannheim: Bibliographisches Institut.

Schmitz-Gunther, T. (ed.). 1999. *Living Spaces: Sustainable Building and Design* [English adaptation by Loren E. Abraham and Thomas A. Fisher]. Cologne, Germany: Konemann Verlagsgesellschaft.

Schneider, D.M; Godschalk, D.R.; Axler, N. 1978. *The Carrying Capacity Concept as a Planning Tool.* Report No. 338. Chicago: American Planning Association.

Schubert, G.H.; Pitcher, J.A. 1973. *A Provisional Tree Seed-zone and Cone-crop Rating System for Arizona and New Mexico.* USDA Forest Service Research Paper RM-105. Ft. Collins, CO: Rocky Mountain Forest and Range Experiment Station.

Schultz, J. 1995. *The Ecozones of the World: The Ecological Divisions of the Geosphere* (transl. from German by I. And D. Jordan). Berlin: Springer-Verlag.

Shirazi, M.A. 1984. Land classification used to select abandoned hazardous waste study sites. *Environmental Management* 8:281–286.

Shreve, F. 1942. The desert vegetation of North America. *Botanical Review* 8:195–246.

Smith, R.L.; Smith, T.M. 2001. *Ecology and Field Biology,* 6th ed. San Francisco: Benjamin Cummings.

Snyder, G. 1996. *A Place in Space: Ethics, Aesthetics, and Watersheds.* Washington, DC: Counterpoint Press.

Soule, J.D.; Piper, J.K. 1992. *Farming in Nature's Legacy: An Ecological Approach to Argiculture.* Washington, DC: Island Press, pp. 133–134.

Soule, M.E.; Terborgh, J. (eds.). 1999. *Continental Conservation: Scientific Foundations of Regional Reserve Networks.* Washington, DC: Island Press.

Spellerberg, I.F.; Sawyer, J.W.D. 1999. *An Introduction to Applied Biogeography.* Cambridge: Cambridge University Press.

Stegner, W. 1992. *Beyond the Hundredth Meridian: John Wesley Powell and the Second Opening of the West.* New York: Penguin Books.

Stein, B.A. 1996. Putting nature on the map. *Nature Conservancy* Jan./Feb.: 24–27.

Stein, B.A.; Kutner, L.S.; Adams, J.S. (eds.). 2000. *Precious Heritage: The Status of Biodiversity in the United States.* New York: Oxford University Press. A joint project of The Nature Conservancy and the Association for Biodiversity Information.

Steiner, F. 1991. *The Living Landscape: An Ecological Approach to Landscape Planning.* New York: McGraw-Hill.

Steiner, J.J.; Poklemba, C.J. 1994. *Lotus corniculatus* classification by seed globulin polypeptides and relationship to accession pedigrees and geographic origin. *Crop Science* 34:255–264.

Stern, J. and M. 1992. *Roadfood.* New York: HarperCollins.

Stilgoe, J.R. 1983. *Metropolitan Corridor: Railroads and the American Scene.* New Haven, CT: Yale University Press.

Stolzenburg, W. [illustrations by Michael McNelly] 1998. When nature draws the map. *Nature Conservancy* Jan./Feb.: 12–23.

Stoms, D. M.. et al. 1998. Gap analysis of the vegetation of the intermountain semi-desert ecoregion. *The Great Basin Naturalist* 58:199–216.

Stuller, J. 1995. Climate is often a matter of inches and a little water. *Smithsonian* 26(9):103–110.

Susanka, S. 1998. *The Not So Big House: A Blueprint for the Way We Really Live.* Newtown, CT: Taunton Press.

Swanson, F.J., et al. 1987. Mass failures and other processes of sediment production in Pacific Northwest landscapes. In *Streamside Management, Forestry and Fisheries Interaction, Proceedings of College of Forest Resources Symposium.* Seattle: University of Washington, pp. 9–38.

Swanson, F.J.; Franklin, J.F.; Sedell, J.R. 1990. Landscape patterns, disturbance, and management in the Pacific Northwest, USA. In I.S. Zonneveld and R.T. Forman, (eds.). *Changing Landscapes: An Ecological Perspective.* New York: Springer-Verlag.

Swanson, F.J.; Kratz, T.K.; Caine, N.; Woodmansee, R.G. 1988. Landform effects on ecosystem patterns and processes. *Bioscience* 38:92–98.

Swanston, D. 1971. Judging impact and damage of timber harvesting to forest soils in mountainous regions of western North America. In *Maintaining Productivity of Soils. Proceedings of the 62d Western Forestry Conference;* 30 November 1971; Portland, OR. Portland, OR: Western Forestry and Conservation Association, pp. 14–20.

Tansley, A.G. 1935. The use and misuse of vegetation terms and concepts. *Ecology* 16:284–307.

Thayer, R.L. 1994. *Gray World, Green Heart: Technology, Nature, and the Sustainable Landscape.* New York: John Wiley & Sons.

Thomas, K.A.; Davis, F.W. 1996. Applications of GAP analysis data in the Mojave Desert of California. In J.M. Scott, T.H. Tear, and F.W. Davis (eds.). *Gap Analysis: A Landscape Approach to Biodiversity Planning.* Bethesda, MD: American Society of Photogrammetry and Remote Sensing, pp. 209–219.

Thomashow, M. 1995. *Ecological Identity.* Cambridge, MA: MIT Press.

Thompson, G.F.; Steiner, F. (eds.). 1997. *Ecological Design and Planning.* New York: John Wiley & Sons.

Thornthwaite, C.W. 1933. The climates of the Earth. *Geographical Review* 23:433–440. With separate map at 1 : 77,000,000.

Thornthwaite, C.W. 1954. Topoclimatology. In *Proceeding of the Toronto Meteorological Conference* Toronto: Royal Meteorological Society, pp. 227–232.

Toffler, A. 1970. *Future Shock.* New York: Random House.

Tosta, N. 1994. Bio-eco-geo-regions. *Geo Info Systems* 4(3):26–28.

Trewartha, G.T. 1968. *An Introduction to Climate,* 4th ed. New York: McGraw-Hill.

Trewartha, G.T.; Robinson, A.H.; Hammond, E.H. 1967. *Physical Elements of Geography,* 5th ed. New York: McGraw-Hill.

Tricart, J.; Cailleux, A. 1972. *Introduction to Climatic Geomorphology* [transl. from French by C.J. Kiewiet de Jonge]. New York: St. Martin's Press.

Tricart, J.; Kiewiet de Jonge, C. 1992. *Ecogeography and Rural Management.* Essex, UK: Longman.

Tufte, E.R. 1990. *Envisioning Information.* Cheshire, CT: Graphics Press.

Turner, M.G.; Gardner, R.H.; O'Neill, R.V. 2001. *Landscape Ecology in Theory and Practice: Pattern and Process.* New York: Springer-Verlag.

Udvardy, M.D.F. 1975. *A Classification of the Biogeographical Provinces of the World.* Occasional Paper No. 18. Morges, Switzerland: International Union for Conservation of Nature and Natural Resources.

USDA Forest Service. 1977. *Silvicultural Activities and Non-point Pollution Abatement: A Cost-effectiveness Analysis Procedure.* Report No. EPA-600/8-77-018. Washington, DC: U.S. Environmental Protection Agency. In cooperation with Environmental Research Laboratory, Athens, GA.

USDA Soil Survey Staff. 1975. *Soil Taxonomy: A Basic System for Making and Interpreting Soils Surveys.* Agricultural Handbook 436. Washington, DC: U.S. Department of Agriculture.

U.S. General Accounting Office. 1994. *Ecosystem Management: Additional Actions Needed to Adequately Test a Promising Approach.* Report No. GAO/RCED-94-111. Washington, DC: U.S General Accounting Office.

U.S. Geological Survey 1979. *Accounting Units of the National Water Data Network.* Washington, DC. Map, scale = 1 : 7,500,000.

U.S. National Arboretum, Agricultural Research Service, Department of Agriculture. 1965. *Plant Hardiness Zone Map* Misc. Publ. No. 814. Washington, DC: USDA Argicultural Service. In cooperation with the American Horticultural Society. Scale = 1 : 7,500,000.

Vale, T.R. 1982. *Plants and People: Vegetation Change in North America.* Washington, DC: Association of American Geographers.

Van der Ryn, S.; Cowan, S. 1996. *Ecological Design.* Washington, DC: Island Press.

Wallace, A.R. 1876. *The Geographic Distribution of Animals.* New York: Harper.

Walter, H.; Box, E. 1976. Global classification of natural terrestrial ecosystems. *Vegetatio* 32:75–81.

Walter, H. 1985. *Vegetation of the Earth and Ecological Systems of the Geobiosphere,* 3rd rev. and enlarged ed. [transl. from German by Owen Muise]. Berlin: Springer-Verlag.

Walter, H.; Harnickell, E.; Mueller-Dombois, D. 1975. *Climate-diagram Maps of the Individual Continents and the Ecological Climatic Regions of the Earth.* Berlin: Springer-Verlag.

Walter, H.; Lieth, H. 1960–1967. *Klimdiagramm Weltatlas.* Jena, Germany: G. Fischer Verlag.

Wann, D. 1994. *Biologic: Designing with Nature to Protect the Environment.* Boulder, CO: Johnson Books.

Webb, W.P. 1931. *The Great Plains.* Waltham, MA: Blaisdell.

Weinstein, G. 1999. *Xeriscape Handbook: A How-to Guide to Natural, Resource-wise Gardening.* Golden, CO: Fulcrum Publishing.

Wells, M. 1994. *Infrastructures: Life Support for the Nation's Circulatory Systems.* Brewster, MA: Underground Art Gallery.

Williams, D.R.; Stewart, S.I. 1998. Sense of place: an elusive concept that is finding a home in ecosystem management. *Journal of Forestry* 96(5):18–23.

Wilson, E.O. 1998. *Consilience: The Unity of Knowledge.* New York: Knopf.

Wilson, J.E. 1999. *Terroir: The Role of Geology, Climate and Culture in the Making of French Wines.* Berkeley: University of California Press.

Wilson, L. 1968. Morphogenetic classification. In R.W. Fairbridge, (ed.). *The Encyclopedia of Geomorphology.* New York: Reinhold, pp. 717–731.

Woodward, J. 2000. *Waterstained Landscapes: Seeing and Shaping Regionally Distinctive Places.* Baltimore, MD: Johns Hopkins University Press.

Young, G.L., et al. 1983. Determining the regional context for landscape planning. *Landscape Planning* 10:269–296.

Zwinger, A. 2000. Human law/natural law: Whose world is this anyway? *Plateau Journal* 4(1):6–19.

Index

About the Author and the Illustrator

ROBERT G. BAILEY (b. 1939, California Coastal Chaparral ecoregion) received his Ph.D. in geography from the University of California, Los Angeles, in 1971. A geographer with the U.S. Forest Service in Fort Collins, Colorado, he was leader of the agency's Ecosystem Management Analysis Center for many years. Currently, he works for the Inventory & Monitoring Institute, where he is in charge of ecoregion studies. He has over three decades of experience working with the theory and practice of ecologically based design and management. He is author of numerous publications on this and related subjects, including two books.

LEV ROPES (b. 1936, Middle Rocky Mountains ecoregion) is a semi-retired consultant, designer, and computer-graphic artist with many years of experience helping people to graphically communicate complex scientific and technical ideas to their audiences. He worked as a limnologist, groundwater hydrologist, and geochemist for the U.S. Geological Survey. He later founded LCT Graphics, Inc., a Denver firm that designed and produced exhibits and presentation materials for science and technology, as well as litigation. He works currently under the name of Guru Graphics.